Glacial Lake Missoula
and Its Humongous Floods

DAVID ALT

Mountain Press Publishing Company
Missoula, Montana
2001

Twelfth Printing, July 2019

All photographs and illustrations by the author unless otherwise credited.

Front cover photograph © 2001 by Mark Alan Wilson, www.floodphotos.com
Dry Falls, Washington

Back cover photograph © 2001 by Ted Wood
Palouse Falls, Washington

Library of Congress Cataloging-in-Publication Data

Alt, David D.
　　Glacial Lake Missoula : and its humongous floods / David Alt.
　　　　p.　cm.
　　Includes bibliographical references (p.).
　　ISBN 978-0-87842-415-3 (alk. paper)
　　　　1. Glacial epoch—Northwest, Pacific. 2. Floods—Northwest,
　　Pacific. 3. Missoula, Lake.
　　I. Title
　　QE697 .A48 2001
　　551.7'92'09795—dc21

　　　　　　　　　　　　　　　　　　　　　　　　　　　　　2001030729

PRINTED IN THE UNITED STATES OF AMERICA

Mountain Press
PUBLISHING COMPANY
P.O. Box 2399 · Missoula, MT 59806 · 406-728-1900
800-234-5308 · info@mtnpress.com
www.mountain-press.com

For Sandra

Contents

Preface

This is the story of an enormous lake that existed during the last ice age, and of the horrifying floods it unleashed when it suddenly drained. I wrote this account of the lake and its floods for people who are not geologists, hoping geologists will also enjoy it.

Glacial Lake Missoula was among the largest lakes ever impounded behind an ice dam. The ice dam broke when the water behind it got deep enough to float it. Each time that happened—several dozen times—the lake dumped a catastrophic flood on eastern Washington and the Columbia River. Those lakes and floods are among the largest of known geologic record. This book follows their story in the fluid logic of gravity, in the direction the waters flowed.

This is also a story of scientists grappling with an emerging scientific controversy. Some handled it well, others miserably as personalities, pride, and outright prejudice superceded scientific evidence. This is not how science should work, but how it often does work. Scientists are fallible human beings, and science is a human pursuit. Those who work on its cutting edge operate in a gray zone where facts often unravel and theories fail to follow their textbook paradigms.

Glacial Lake Missoula and its great floods left their tracks in many places in four states. It would be impossible to visit all of them in just one book, so I selected a few of the most spectacular, the most accessible, the most typical, and the most perplexing. The scientific literature describes most of these sites. I relied heavily on these sources, and thank their authors, who invested so much time and energy in finding the pieces of this story. I also visited these sites, some of them many times, and in places include some of my own observations as well.

I first intended to cite all the references and their authors in the text, but that soon proved impossibly cumbersome in a book addressed primarily to a lay audience. So I tried to simplify matters by using informal citations of only the most important references. But that seemed to slight too many people who had done so much good work. So I began to compile a bibliography. But it soon became obvious that the bibliography would never be complete. So I settled for a partial bibliography.

Most of the literature references are available only in the libraries of research universities, many only through interlibrary loan. Research libraries within the area of Glacial Lake Missoula and its floods include those at the University of Montana in Missoula, Washington State University in Pullman, the University of Idaho in Moscow, Portland State University in Portland, Oregon State University in Corvallis, and the University of Oregon in Eugene.

I thank the good people at Mountain Press for their excellent help in editing and designing this volume.

A number of people read versions of the manuscript as it slowly evolved and offered many helpful suggestions. I especially thank Victor R. Baker, who thoroughly reviewed the entire manuscript and made many perceptive suggestions. Tom Thompson, Jim Sears, Jeff Silkwood, Don Hyndman, Steve Sheriff, Annie Gallatly, and Emilie Reagan all offered many stimulating suggestions. Perri Knize did an exceptionally enthusiastic job of copyediting. I thank these people for having rescued me from any number of potentially embarrassing mistakes.

Jeff Silkwood and the U.S. Forest Service generously made available early versions of their elegant digital map of Glacial Lake Missoula and the areas it flooded. The published version is available through the U.S. Forest Service at a nominal price. Don Hyndman and John Rimel provided a number of excellent photographs. Finally, a plea. Many of the relics of Glacial Lake Missoula and its catastrophic floods are world-class features. We have already sacrificed too many of them to plows, roads, and urban sprawl. The best of those that remain intact should become the focal points of parks, or at least of roadside signs. Our children and grandchildren will surely thank us if we preserve as much as we can of this unique heritage.

INTRODUCTION
A Glacial Lake, the Fertile Palouse Hills, the Barren Scablands

THE FARMERS WHO SETTLED in eastern Washington in the late 1800s knew a few things about farmland. They were mightily impressed with the crops they raised in the rolling Palouse Hills. That was wonderful farmland, a softly rounded landscape deeply upholstered in richly fertile soils, mostly silt that blew in from the southwest and piled into enormous dunes. The Palouse silt makes excellent soil.

But set within the rich Palouse country those farmers found tracts of scabby outcrops of black basalt, broad expanses of raw gravel, and dry stream channels, which people in Washington call coulees. Those harsh areas were nearly worthless as farmland. The farmers called them scabland.

Some of the scabland coulees are very large, and many of them come in swarms that make complexly braided patterns on the map. The coulees and swarms of coulees in the scabland look as though they carried enormous volumes of water at some past time. That time could not have been all that long ago, if we can judge from the generally fresh appearance of the coulees. Some scabland channels were ordinary stream valleys before floods scoured them. Others cross divides as much as 300 feet above the present streams. The torrents of water that rushed through those channels stripped the pale Palouse silt in many places, then carved deep into the black basalt bedrock. Those parts of the scablands look from the air like dark streaks scribbled with giant crayons across the generally pale countryside.

1

Scabland channel near Sprague, Washington —D. W. Hyndman photo

From the ground you see coulees that hold pothole ponds and small lakes almost beyond counting. Broad expanses of coarse gravel spread across vast areas. Scabby outcrops of basalt stand amid groves of scrubby pine trees.

No one knows exactly when the most recent ice age started, maybe sometime around 40,000 years ago. But evidence of many kinds and from many places clearly shows that the ice finally reached its maximum extent about 15,000 years ago. By that time, ice so deeply filled the valleys of British Columbia that only the tops of the highest mountains rose above it. The southern fringe of that regional ice cover encroached into the mountain valleys of northern Washington and Idaho and northwestern Montana. Enormous lobes of ice pushed farther south as they filled the broad Flathead and Swan Valleys of Montana, the Purcell Valley of Idaho, and the Okanogan Valley of Washington.

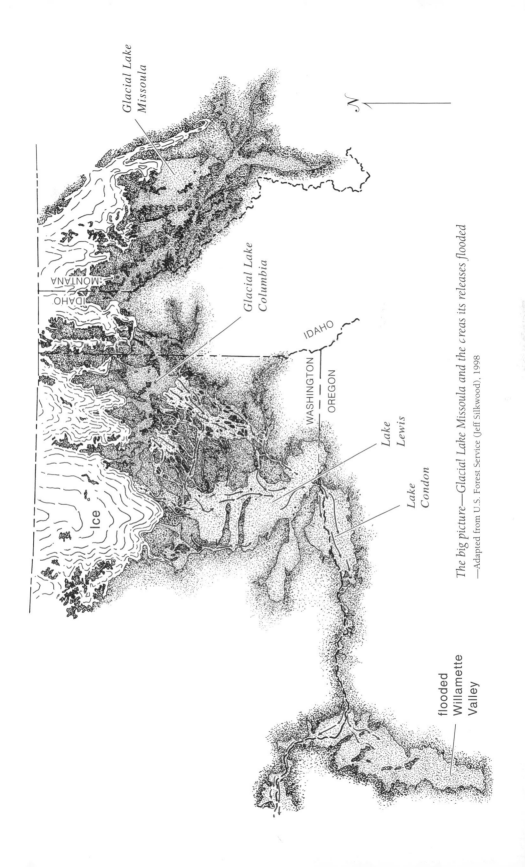

The big picture—Glacial Lake Missoula and the areas its releases flooded
—Adapted from U.S. Forest Service (Jeff Silkwood), 1998

Glacial Lake
Missoula

Glacial Lake
Columbia

Lake
Lewis

Lake
Condon

flooded
Willamette
Valley

MONTANA
IDAHO

IDAHO
WASHINGTON
OREGON

Ice

N

As the glaciers of the most recent ice age reached their maximum spread about 15,000 years ago, the ice pouring south down the Purcell Valley of northern Idaho crossed the valley of the Clark Fork River at the present site of Lake Pend Oreille, just south of Sandpoint. Lobes of that glacier spread west down the present valley of the Pend Oreille River and east up the valley of the Clark Fork River, almost to the Montana line. The mass of ice filling those valleys was at times more than 30 miles across. It dammed the Clark Fork River and impounded Glacial Lake Missoula.

The lake had no outlet across the ridges that enclose the drainage basin of the Clark Fork River, so the water rose from one year to the next until it was finally deep enough to float its ice dam. The dam broke and the lake suddenly drained, its waters sweeping broadly— and catastrophically—across eastern Washington and down the Co- lumbia River to the Pacific Ocean. But the relentlessly advancing ice soon established a new ice dam that impounded a new version of Glacial Lake Missoula, which in due course floated its ice dam and released another great flood. This happened several dozen times.

The great Cascade volcanoes provide an approximate time line for the floods. Age dates of layers of volcanic ash sandwiched be- tween the older flood deposits in eastern Washington show that the first of those great floods swept across the Northwest sometime around 15,000 years ago. The glaciers of the last ice age were near their far- thest reach and greatest thickness then and moving rapidly.

Then the climate abruptly changed, and the great glaciers began to melt so rapidly that they would be gone within about 3,000 years. That really is rapid, considering that the ice had been accumulating for more than 30,000 years and was thousands of feet thick in many places. Age dates on layers of volcanic ash sandwiched between the younger flood deposits show that the last Glacial Lake Missoula floods happened sometime around 13,000 years ago. That was probably about when the rapidly thinning ice in the Purcell Trench finally stopped moving and could no longer advance across the Clark Fork River.

In 1978, P. C. Patton and his colleagues summarized evidence they had seen of torrential floods that passed through the scablands long before the most recent ice age. They found exposures of gravel that appeared to have been dumped from fast flood waters. This gravel had a thick cover of windblown silt on which a deep topsoil had developed. On top of that topsoil, the floods of the most recent ice age dumped their gravel. In some places, the buried gravel contains deeply weathered and thoroughly rotten cobbles of basalt.

The accumulation of the windblown silt surely took many years, probably thousands of years. And it also took a very long time to weather the basalt cobbles within the silt. The deposits of the most recent ice age have not acquired a similar cover of windblown silt and soil, and the chunks of basalt within them have not weathered much in the 13,000 or so years since they were dumped.

So the older flood gravels are very old indeed, but it is not clear how old. Neither is it clear whether they date from a single earlier ice age, or from several. But it does seem clear that the several dozen great floods of the most recent ice age were not entirely responsible for shaping the scablands.

Dozens of Glacial Lake Missoula shorelines on Mount Jumbo, on the east side of the Missoula Valley —D. W. Hyndman photo

1

TRULY AN INLAND SEA
Glacial Lake Missoula

T. C. CHAMBERLAIN was a geology professor at Northwestern University during the academic year. But it seems that he, like most university professors, needed a summer job to make ends meet. The late summer of 1886 found him in the Mission Valley of western Montana, where he was winding up a long summer of field work spent mapping glacial features for the U.S. Geological Survey. In his final report, he vaguely mentioned having seen faint "watermarks," as he called them, that looked like "giant musical staves" wrapping around the hills. He concluded that a glacial lake had once flooded the valley. That, so far as I know, is the earliest published reference to the old shorelines of Glacial Lake Missoula.

Chamberlain made an easy observation and arrived at an easy conclusion. Old shorelines are unique in the landscapes of the world in being perfectly horizontal and perfectly parallel. And the geologists of Chamberlain's time had all read a classic description of a famous pair of glacial lake shorelines in Scotland, the "parallel roads of Glen Roy." So while Chamberlain easily identified the old shorelines, they were tangential to his main duties. The summer was almost over, so he did not pursue the subject. It waited for J. T. Pardee.

Joseph T. Pardee was born in 1871 and grew up in western Montana mining towns. From 1889 to 1891, he studied chemistry and mining at the University of California at Berkeley. Then he returned to western Montana, opened an assay office in Philipsburg, and

Joseph T. Pardee
—Courtesy Geological
Society of America
(undated photograph)

operated a small placer mine nearby. When he married in 1904, he sold the assay office and the mine and bought a small farm near Stevensville in the Bitterroot Valley. There he spent much of his spare time studying geology books—in those days, geology was still more a calling than a major in college. Many geologists learned their trade on the job.

Pardee passed the civil service examination in Missoula in 1908 and went to work for the U.S. Geological Survey in 1909. His first task as a professional geologist was to investigate Glacial Lake Missoula. We can safely assume that for years he had wondered about the old shorelines on the grassy hills around the Missoula and Bitterroot Valleys. He probably also knew that Chamberlain had mentioned old shorelines in the Mission Valley.

Nature intends all surface water to run down a stream to the ocean. All lakes pose a geologic problem because they exist only if some sort of dam impounds the drainage. So the first question in a geologic investigation of any lake, ancient or modern, is the location and nature of the impounding dam.

Pardee undoubtedly spent a good part of 1909 following the shore-lines of Glacial Lake Missoula down the long slope of the Clark Fork River to the panhandle of northern Idaho, where they disappeared. That, evidently, was where the dam had been. But today the river flows into Pend Oreille Lake through a broad valley that contains no hint of a dam. Pardee likely had also read about the ice dam respon-sible for the "parallel roads of Glen Roy," and he probably had a precedent in mind.

Pardee certainly understood that a sea of regional ice had deeply filled the valleys of British Columbia during the most recent ice age. He concluded that a great lobe of that ice had come south down the Purcell Valley, dammed the Clark Fork River, and impounded Glacial Lake Missoula. Of course, the dam disappeared without a trace when it melted near the end of the most recent ice age. Pend Oreille Lake now floods much of its site.

Having identified its vanished dam, Pardee went on to provide some of the vital statistics of Glacial Lake Missoula. He estimated that its surface covered approximately 2,900 square miles when at its highest shoreline, at an elevation of about 4,150 feet above sea level. At that level, it held approximately 500 cubic miles of water, about half the volume of Lake Michigan. It was, truly, an inland sea. At its maximum filling, Glacial Lake Missoula was almost 2,000 feet deep at the ice dam and about 950 feet deep in the Missoula Valley.

It is easy to reconstruct a map of Glacial Lake Missoula filled to the elevation of any of its shorelines. Just trace the contour for that eleva-tion on a topographic map, and there it is, that simply. All those reconstructions of the old shorelines look like intricate pieces of an-tique lace because the contours faithfully trace every little tributary.

Reconstructing the volume of Glacial Lake Missoula is easy in prin-ciple, but it was tedious work in the days before computers. Pardee probably used an instrument called a planimeter to measure the areas enclosed within the topographic contours, then estimated the volume of water contained between them as 500 cubic miles. In 1996, Jeff Silkwood of the U.S. Forest Service used a computer to extract from the latest digital elevation data a volume of about 530 cubic miles. I assume the modern figure is an improvement, but marvel at how little it differs from Pardee's original estimate.

Glacial Lake Missoula flooded all the mountain valleys of the Clark Fork drainage in western Montana. When it stood at its highest shoreline, Glacial Lake Missoula reached south beyond Darby in the Bitterroot Valley. It filled the Clark Fork Valley east a bit beyond Garrison. It filled the Mission Valley to Polson, where it lapped against the southern end of the ice that had flowed south from Canada and filled the Flathead Valley. And it flooded the Potomac and Clearwater Valleys almost as far north as the Canadian ice that filled the Seeley-Swan Valley. It fingered into every tributary and tributary of a tributary.

When Pardee described it, Glacial Lake Missoula was the largest lake known to have existed behind an ice dam, anywhere. It retained that distinction into the 1990s. Then V. R. Baker, a distinguished hydrologist and prominent student of the Glacial Lake Missoula floods, and two colleagues discovered an apparently larger example in the Altay Mountains of Siberia. Others probably exist, quite possibly in the northward drainage of Glacial Lake Aggasiz, which flooded a large area of eastern North Dakota, western Minnesota, and southern Manitoba.

When Pardee published his account of Glacial Lake Missoula in 1910, it must have seemed that he had told us more than anyone would ever need to know about that subject. He went on to a distinguished career in which he tackled a long series of problems that had nothing to do with glacial lakes. He especially distinguished himself through his studies of deposits of manganese and phosphate and the structure of the northern Rocky Mountains. He could not have dreamed in 1910 that he would crown that career with a second study of Glacial Lake Missoula in the midst of a great controversy.

2

A WINDOW INTO THE PAST
A Visit to Glacial Lake Missoula

GLACIAL LAKE MISSOULA existed long ago and much farther south than modern glacial ice dams and their lakes. Nevertheless, it probably resembled the modern glacial lakes enough to make them a useful guide for a visit to the vanished Pleistocene past.

The great continental ice sheets were melting very fast in the time of Glacial Lake Missoula, shedding great torrents of meltwater every summer, raising sea level an average of approximately 100 feet every 1,000 years. Meltwater swept enormous volumes of sediment just freed from the melting ice into every low place it reached. The glacial lakes got their fair share. All glacial lakes, ancient and modern, deposit distinctive sediments unlike those laid down in any other setting.

Moving glaciers grind together the rocks frozen within them, pulverizing them into powdery rock flour, which is generally very pale gray, almost white. The warmth of the long summer afternoons of the most recent ice age sent great freshets of glacial meltwater, milky white with suspended rock flour, into Glacial Lake Missoula. Rock flour settled through the murky water and deposited as a layer of pale sediment on the floor of the lake. Those pale summer layers varied in thickness from a fraction of an inch to as much as 2 inches, or even more.

Meanwhile, the suspended rock flour enriched the lake in dissolved mineral nutrients, making an extremely fertile environment for algae and the long food chain of animals that depend on them.

Algae bloomed in the sunlit shallows of summer, making a rich organic soup. Then the cooling weather of fall stopped the glacial melt-water and cut off the supply of rock flour. After the winter freeze killed summer's algae and microscopic life, their remains slowly settled into a layer of sediment dark with organic matter, the color of dark chocolate, on the lake floor. Those winter layers also range in thickness from a fraction of an inch to as much as 2 inches, or occasionally more.

Geologists call the layers of light and dark sediment laid down on the floors of glacial lakes varves. A lake's varves are its archives, a record of its summers and winters. You see the records of warm or cool summers in pale layers that are thicker or thinner than most. Find a thin streak of pale rock flour in a dark winter layer, and you see the lake's memory of a midwinter thaw.

Very small amounts of rock flour color a lake a startling greenish blue. Glacial Lake Missoula surely became a splendid and brilliant greenish blue as the last of the summer rock flour settled and the larch trees blazed yellow in the deepening chill of the coming winter. Then the lake lay dark under the gray winter sky until ice began to spread from its shores. Such a deep lake probably did not freeze over, but its open surface surely floated slabs of pack ice. Icebergs also drifted on parts of the lake. The masses of floating ice certainly stilled the waves, preventing much heavy surf along the shores.

Icebergs are big pieces of ice that break off the floating end of a glacier, then drift across open water. Only the parts of Glacial Lake Missoula that submerged the lower ends of glaciers could have launched icebergs, and those places were few. Icebergs were scarce in Glacial Lake Missoula. They could have existed only when the lake was at its higher fillings, deep enough to float the lower ends of glaciers that emerged from valleys in the Mission Mountains, the Bitterroot Mountains, and possibly the Rattlesnake Mountains.

The deeper parts of glaciers commonly contain large amounts of sediment in all sizes, from silt to boulders the size of cars, or larger. Icebergs drop that sediment as they drift or when they run aground and finally melt. A close group of boulders of different kinds of rocks is the characteristic signature of a grounded iceberg that melted long ago. Floating pack ice that forms as a lake freezes typically contains very little sediment, and it is small stuff—no boulders.

Icebergs could hardly run aground and stay grounded for very long during the years the lake level was rising. A few probably did while the lake was draining, but even then, the flood probably swept most of them along. So boulders that icebergs floated into places remote from their homes are rare in the valleys of western Montana. They are more common in some areas of Washington and Oregon, where icebergs rode in on the great floods, instead of out with the great emptyings.

It is surprising that no one has found fossils in Glacial Lake Missoula sediments—no petrified wood, no leaves, no bones. We have no tangible record of the plants and animals that lived in Glacial Lake Missoula, or around its shores. But we can reasonably speculate.

The weather was very wet during the most recent ice age—plenty of rain and snow. And most geologists and biologists who have given the subject much thought agree that most of the plants that grew around Glacial Lake Missoula and most of the menagerie of animals that prowled its shores were the same species that now live in western Montana. But we have lost a few.

Mastodons, for example, were common during the most recent ice age, if we can judge from the abundance of their bones in river and glacial sediments elsewhere. They looked almost like modern elephants except that they were much larger and clothed in long hair. It's easy to imagine them trumpeting their greetings to the morning along the shores of Glacial Lake Missoula. Beavers the size of grizzly bears lived then. So did bison much larger and with much longer horns than those that so recently roamed in great herds across the high plains. Evidence elsewhere in the region makes it seem reasonable to imagine these creatures populating the shores of Glacial Lake Missoula, even if we do not have their bones.

It is especially surprising to find evidence of a great lake with no sign of fish, not even a few scattered scales. Why no fish fossils? Perhaps no fish. The summer murk of suspended rock flour probably made the lake a poor habitat for most of the kinds of fish native to western Montana. And the lake's sudden drainings surely flushed any fish that may have been around.

It would be nice to know that people were around to see Glacial Lake Missoula and its humongous floods. I would like to think that those spectacles were not entirely wasted on hairy mastodons, giant beavers, oversized bison. That may not be a forlorn hope.

Archaeologists have good evidence that people lived in North America before the end of the last ice age, perhaps thousands of years before. Some of them may have known the lake and watched its great floods. But the archaeologists have not yet dug up any direct evidence that would place people in the Pacific Northwest during that time. Still, we can imagine them in those landscapes, witness to an enormous lake and the horrendous floods it unleashed.

The scars on the landscape tell us something of what people might have seen as Glacial Lake Missoula filled the valleys of the Clark Fork River in western Montana, then emptied in a few days as it dumped a great flood across eastern Washington and down the Columbia River to the Pacific Ocean. We can only imagine what they might have felt.

3

A Heretic in the Scablands
J Harlen Bretz and His Humongous Flood

J Harlen Bretz was the very model of a careful and thoughtful scientist, always meticulous in his close attention to the evidence at hand and the question of how best to interpret it. Photographs taken during his middle years show a man of middling size, compactly built, adorned with a neatly trimmed mustache. Some inlcude a sweaty old hat that looks just right on a man so passionately devoted to geologic field work. And he may have been a bit of a firebrand.

Bretz was born in Michigan in 1881 and grew up there. The J without a period was his first name, not an initial. When the time came, he went to the University of Chicago, where he majored in biology. He started his career teaching biology at a high school in Seattle, where he became so interested in the glaciated landscapes of the Puget Sound region that he decided to switch to geology. So he returned to the University of Chicago, where he earned a doctorate in geology in 1913. He taught for one year at the University of Washington, then returned to the University of Chicago, where he was a faculty member for the rest of his exceedingly long career.

Somewhere along the way, Bretz developed an abiding interest in the scablands of eastern Washington. He always called them the Channelled Scablands, and many geologists follow his example. I do not because many of the scablands do not include channels. Call them what you will, those bizarre landscapes summoned him back summer after summer for long and strenuous campaigns of

J Harlen Bretz
—Courtesy Department
of Special Collections,
The University of
Chicago Library
(undated photograph)

field work with small groups of students. This was long before geologists used aerial photos, so Bretz patiently walked the dry scablands, slowly generalizing his detailed local observations into a sweeping regional aerial perspective.

Bretz was 41 years old in the summer of 1923 and back for his last season of field work in the scablands. He again prowled the dry channels with his students, while the endless fields of Palouse wheat ripened in the baking summer sun. His endless walking and thinking were finally crystallizing ideas that would soon ignite enough controversy to satisfy almost any firebrand.

J Harlen Bretz presented his first paper about the scablands in the fall of 1923 at the annual meeting of the Geological Society of America. The published version painstakingly describes the scablands, complete with photographs and maps that methodically take the reader from Spokane to the Pasco Basin. Having presented that great mass of evidence, he interpreted the scablands as the result of glacial erosion and of torrential flows of water from melting glacial ice. After his presentation, several geologists, some eminent and others destined to become so, rose in the audience to congratulate him for having made a splendid contribution to knowledge.

At another meeting about two months later, Bretz again presented essentially the same evidence in a second paper, but by then he had drawn radically different conclusions. In an extraordinary reversal, he boldly proposed that an enormous flood had eroded the scablands in a very brief time, perhaps a matter of days. Once again, geologists rose in the audience to comment, but this time they roundly denounced his conclusions and the thinking that had led to them.

An old professor of my undergraduate days had been a graduate student sitting in the audience when Bretz read that second paper. The professor was a man of many talents who did a hilarious impersonation of Bretz pounding on the podium with both fists and stomping on the floor as he used vivid language and gestures to convey his idea of a catastrophic flood to his horrified audience. Bretz rarely left his audience in any doubt of his meaning.

The geologists in that second audience were aghast in the same way that a roomful of physicists would be upon hearing a colleague explain how he had made a perpetual motion machine out of old popsicle sticks. Physicists had all learned very early of the futility of perpetual motion machines, and no properly educated geologist was supposed to traffic in catastrophes of any sort.

Scientific geology began during the 1790s, when early geologists began to explain rocks and landscapes as the result of slow processes and weak forces operating over long periods of time. That was a radical departure from their earlier practice of attributing geologic events to great and sudden catastrophes. The geologists who listened as Bretz read his second paper of 1923 were shocked to hear him invoke a sudden catastrophe to explain the scablands landscapes of eastern Washington. In their view, this was a reversion to the unscientific thinking of some 125 years before. To this day, most geologists consider it nothing less than heresy to invoke a catastrophic explanation for a geologic event.

So Bretz stepped off the end of a very long limb when he suggested that a great flood had eroded the scablands, and he knew it. He was a sophisticated man, no blundering innocent. That second paper on the scablands made him a pariah among geologists, an outcast from the politer precincts of their society. If scientific controversies were settled by a vote, Bretz would have lost immediately, crushed under a landslide of outraged opinion. But being crushed was not part of his personal style. He would continue to defend his ideas for almost sixty more years.

Whatever else it may be, science is not democratic. Although they may prevail for many years, votes, emotions, and dogmas do not finally settle scientific controversies. The only way to do that is with evidence, and Bretz had evidence, mountains of evidence, many different kinds of evidence. He had walked through those scablands, observing every detail, gathering every bit of evidence he could find.

None of Bretz's principal detractors were deeply familar with the scablands. Indeed, some had never seen them and would continue to criticize his ideas for years without bothering to visit the scablands. But they remembered the basic principles they had learned as college freshman. They were quite certain that Bretz was wrong simply because he had invoked a catastrophic event to explain what he had seen. They did not feel any need to consider the evidence. These people knew the answer before they heard the question.

Very few scientists in any discipline come to daring new conclusions as they consider the evidence. Most go through life believing what they learned in college, resenting challenges to their settled beliefs, and disliking those who present them. They muddle their way through their problems in a fog of confusion just like people who are not scientists. That may not be how science should be done, but that is how it very often is done.

Bretz was invited to present another paper about the scablands in 1927, at a meeting of the Geological Society of Washington, D.C. A cabal of his more prominent detractors converted the occasion into what they called a debate, but it was more like an ambush. Some who were there described it as a lynching. Several of the prominent geologists in the audience denounced Bretz's ideas in terms so abrasive they were personal insults. It was, by all accounts, a shameful display, especially considering that some of the most vocal detractors still had not visited the scablands and had no personal knowledge of those extraordinary landscapes. They based their arguments entirely on the received gospel of slow processes, weak forces, and plenty of time.

Several of Bretz's detractors proposed competing scablands theories of their own, none of which explained the evidence. All insisted that the explanation for the extraordinary landscapes of the scablands must somehow lie in the slow operation of ordinary processes. If some of their alternative theories were downright batty, that further illustrates the difficulty of fitting the scablands into the context of conventional geology.

The tempest in the dainty little teapot of academic geology continued after the great debate in Washington, but with little further input from Bretz. He had other fish to fry, mainly a long investigation of the limestone caverns in the Ozark Mountains of southern Missouri. That work led to a number of small controversies, none of them very inflammatory.

Bretz did return for one last season in the scablands in the summer of 1952 in the company of several geologists who were sympathetic to his ideas—their number was slowly growing. The group added many details to his geologic picture. And they found new evidence that persuaded Bretz to argue that more than one great flood had scoured the region—at least seven, perhaps as many as nine.

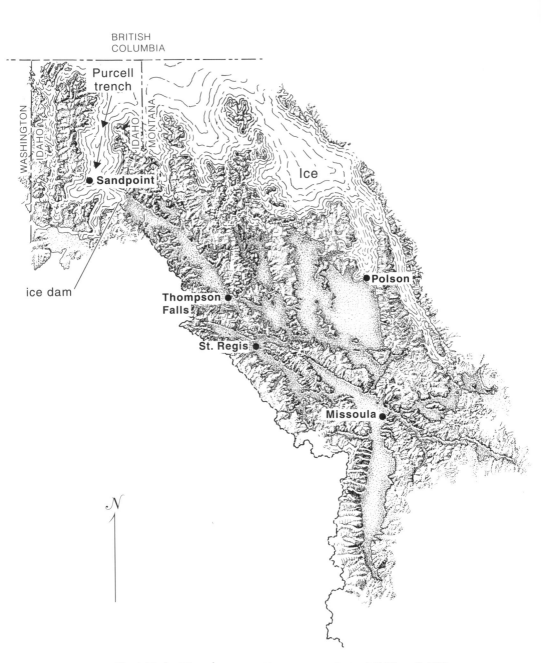

Glacial Lake Missoula —Adapted from U.S. Forest Service (Jeff Silkwood), 1998

WHERE DID THE
WATER COME FROM?
J. T. Pardee Returns to Glacial Lake Missoula

WHEN J HARLEN BRETZ first proposed his great flood, he could not say where the water had come from. He pointed to the enormous expanse of glaciated country to the north and vaguely suggested that the water had come from somewhere up there, somehow. Perhaps a brief interlude of much warmer weather melted an enormous amount of ice. Maybe a volcano erupted beneath the ice. It was a puzzle.

People need not understand everything they know. It is perfectly proper in scientific discussions to recognize that a phenomenon exists without being able to explain it. We do not doubt that birds migrate, even if we do not understand how they find their way. And we know that robins catch worms, even if we are not sure whether they find them by sight or by sound. Bretz did not need to know where the water came from before he could propose his theory of a catastrophic flood. But it would have helped his cause, made his theory more convincing.

J. T. Pardee, who had published the first paper on Glacial Lake Missoula in 1910, confided to colleagues at the 1927 meeting in Washington that he believed he knew where the great flood came from. He had also studied the scablands and published a brief discussion of them in 1922, so his view of the subject was more comprehensive than that of any of Bretz's detractors. None of them were familiar with both the scablands and Glacial Lake Missoula, and some were familiar with neither.

So the idea that the flood had come from Glacial Lake Missoula was in the air by 1927, but no one had seen any direct evidence. J Harlen Bretz did no field work in western Montana and was never eager to offer opinions without strong supporting evidence. He liked to know what he was talking about. Nevertheless, some of his colleagues believed that Bretz was convinced by about 1928 that Glacial Lake Missoula was the probable source of the great flood. But he did not pursue the subject.

J. T. Pardee finally returned to Glacial Lake Missoula in the late 1930s, no doubt with the origin of the great floods in mind. In 1940, the year he retired, he presented a paper about unusual currents in Glacial Lake Missoula. It was published in 1942 as his last major professional work, ending his long career as he began it with a study of Glacial Lake Missoula. I consider that final paper his finest work, full of keen observations and bold insights. Its influence has been profound and lasting.

In that paper of 1942, Pardee described in detail the evidence of catastrophic emptying he found at many sites in the area of Glacial Lake Missoula. He used many different lines of evidence to show that currents of extraordinary power scoured the floor of Glacial Lake Missoula while it emptied. He concluded that the lake drained after the ice dam in northern Idaho suddenly failed. He left it to others to infer for themselves that Glacial Lake Missoula was the source of the flood that scoured the scablands. That was almost glaringly obvious, except to those who continued to insist that no such flood could possibly have happened.

J Harlen Bretz had his water source.

HARD AND
DISTINCTIVE BEDROCK
Belt Rocks

EVEN THOUGH THIS BOOK is not about bedrock, we must pause for a few moments to give the bedrock of Glacial Lake Missoula some thought. The interaction between strong drainage currents and bedrock dictated the shapes of many landscapes in narrow canyons, where the currents were most powerfully erosive. And the bedrock of Glacial Lake Missoula is so distinctive that stray pieces of it are instantly recognizable in settings as remote as the Willamette Valley of western Oregon. Knowing just a little about the hard bedrock of western Montana can solve a good many geologic puzzles.

Except for the Bitterroot Mountains, which consist mainly of granite, the hard bedrock in the Clark Fork drainage of western Montana belongs to a thick stack of layered sedimentary rocks that geologists call Belt rocks. Their formations pile to an unknown depth, considerably more than a dozen miles in some areas. They were deposited, mainly as sands and muds, approximately 1 billion years ago, give or take a few hundred million years. That makes them approximately twice as old as the oldest known animal fossils. Belt rocks are peculiar in containing microscopic crystals of metamorphic minerals that grew at a time when the rocks were hot, probably a billion or so years ago, when almost everything was hotter than now. Belt rocks still look like fairly ordinary mudstones and sandstones, but their

*A rippled
bedding surface
in purple
mudstone, Belt
rock*

content of metamorphic minerals makes them much harder and more durable than ordinary mudstones and sandstones.

The Belt formations contain hundreds of sills of black igneous rock sandwiched between the layers of sedimentary rock. All are diabase, a black rock that differs from basalt only in its texture. Most of the sills are several hundred feet thick. No other rock formations in the area of Glacial Lake Missoula and its floods contain diabase sills, so we know the source of any stray pieces.

Most of the Belt rocks in the western part of the area of Glacial Lake Missoula are mudstones and sandstones in somber shades of medium to dark gray. Many contain the iron sulfide mineral pyrite, which spreads a rusty stain across most weathered surfaces. Farther east, the Belt sandstones and mudstones take on shades of green, red, pink, and yellow. Many come complete with a variety of sedimentary features such as little ripples and intricate patterns of thin layers, all exquisitely preserved. Most people find these colorful rocks full of interesting details and very easy to enjoy. Their colorful pebbles make lovely streambeds.

No outcrops of Belt rocks or diabase exist anywhere west of the northern Rocky Mountains. But chunks of these distinctive rocks litter the entire path of the Glacial Lake Missoula floods, all the way to the Pacific Ocean. The great floods imported all of them—the larger ones frozen in icebergs, the smaller ones simply carried in the deluge. A piece of Belt rock stranded in a field in the Willamette Valley is at least 500 miles from its bedrock home.

THE ONCE AND
REPEATING ICE DAM
Several Dozen Lakes

ICE WEIGHS ABOUT 10 percent less than an equal volume of water, so a lake floats an ice dam when its water level rises higher than nine-tenths of the height of the dam. The ice dam floats in the lake for precisely the same reason that an ice cube floats in a glass of lemonade. But the ice dam breaks as it floats, and it washes out as great icebergs in the sudden flush of the draining lake. Several glacially dammed lakes in Alaska and northern British Columbia drain every year, usually in summer when a fast snowmelt rapidly raises the level of the lake. The ice dam that held Glacial Lake Missoula probably floated and broke during the summer for the same reason, but not every year.

Even though it was very large, as much as 30 miles across, the ice dam was just a tiny southern lobe of the vast sea of regional ice that filled the valleys of British Columbia. After the ice dam washed out, ice continued to move down the Purcell Trench to establish a new dam, probably within a few decades. Then the new lake would fill until it was deep enough to float the new ice dam, and so on, and on. The wetter climates of the ice ages and the extremely fast melting in the final stage of the most recent ice age certainly maintained a much larger version of the Clark Fork River than the one we now know. It surely filled those lakes much more quickly than the modern river could, but we have no way of knowing how much more quickly.

Glacial Lake Missoula could not have floated its ice dam if it had overflowed through a low pass in any of its surrounding drainage divides. If that had happened, the overflow would have established a stable lake surface at the level of the pass. And our only mementos of Glacial Lake Missoula would now be a single shoreline deeply etched in the hillsides and deep deposits of lake sediment on the valley floors—no multiple shorelines, no signs of sudden drainages or catastrophic floods.

In 1971, R. L. Chambers, who was then a graduate student in geology, and I studied a large roadcut in Glacial Lake Missoula sediment beside Interstate 90, just west of the Ninemile exit, about 20 miles west of Missoula. We expected to find a thick sequence of glacial lake sediments and planned to count and study their pairs of light and dark layers that record the summers and winters of the waning years of the most recent ice age. But we soon discovered that the situation was much more complex than that, and far more informative.

The roadcut beside Interstate 90. The pale bands are river silts; the dark bands are sequences of glacial lake sediments with varves. —D. W. Hyndman photo

Instead of one long sequence of light and dark glacial lake varves, we found thirty-six fairly short sequences, with layers of complexly bedded silt sandwiched between them. We interpreted the varved sequences as a record of times when the lake existed. And we became convinced that the complexly bedded layers of silt recorded a time when the lake was empty and the Clark Fork River was depositing sediment on the site—the river is hardly more than a stone's throw away. We concluded that the lake had filled and emptied at least thirty-six times. And we counted a total of just under one thousand pairs of light and dark layers in the thirty-six sections of lake sediments. Those represented the years of Glacial Lake Missoula.

When we counted the varves between the layers of river sediment, we found that the sequence of lake sediments on the bottom of the stack contained a record of 58 years and that each successive sequence above it recorded fewer years than the one beneath. The shortest interval, nine years, was at the top of the stack. Evidently, the ice dam held for fewer years each time the lake refilled. I believe that was because the ice in the Purcell Trench was thinning and therefore floating in shallower water each time. I think we were seeing the great melting at the end of the most recent ice age.

Because Glacial Lake Missoula had no outlet, its level would continue to rise as long as the ice dam lasted. If we assume that the flow of the Clark Fork River remained essentially constant during the years of Glacial Lake Missoula, we can suppose that the oldest sequence of layered lake sediments corresponds to the highest shoreline. Since each successive filling lasted fewer years than the one before, the shorelines presumably become younger as they step down the mountain. All the shorelines are so faint that a bit of wave erosion would erase them. That makes it hard to imagine that the lake level could rise above the level of an earlier shoreline without destroying it.

If each successive filling of Glacial Lake Missoula involved less water than the one before, we would expect to see that pattern expressed in any flood deposits downstream. As we shall see, the oldest flood deposits in eastern Washington are the largest, and they become progressively thinner upward. It seems reasonable to interpret that pattern as evidence that the earliest floods were the greatest and each successive one was smaller than the one before.

Since the lake refilled to a lower level each time, a band of essentially barren ground must always have existed between the water and the trees. It probably looked like the bare slope above a reservoir drawn low after a long dry spell. It seems unlikely that such a lakeshore could have supported the usual assemblage of shoreline plants or the animals that live among them.

The aftermath of the calamitous drainings of Glacial Lake Missoula was a dreadful mess that spanned the six hundred miles from the mountain valleys of western Montana to the Pacific Ocean. Narrow valley walls were scoured to bare bedrock cliffs, no doubt with stray rocks clattering down them. Vast and watery expanses of freshly deposited mud, sand, gravel, and boulders spread across the floors of the broader valleys. The floods left the steep mountain slopes of western Montana and of the Columbia Gorge saturated with water. Suddenly deprived of buoyant support as the lake drained, the hillsides shed landslides and mudflows.

Plants eagerly colonize almost any kind of disturbed ground. So we can imagine the landscape greening as the weedier sorts sprouted in cracks in the cliffs along the flood's path and in the barren deposits of lake sediment and raw gravel. Within a few years, the plants softened the hard edges of devastation, restoring the raw land to a gentler appearance and more productive condition.

Meanwhile, ice continued to move south down the Purcell Valley as though nothing had happened. It once again crossed the valley of the Clark Fork River and established a new dam where the earlier one had washed out. The varve pairs that R. L. Chambers and I found accounted for nearly 1,000 years in thirty-six fillings of the lake. That leaves about 2,000 years in which the site was not flooded. A brief exercise in long division shows that the average interval between fillings of the lake was approximately 50 years.

The earlier fillings of Glacial Lake Missoula came at a time when the glaciers of the most recent ice age were near their maximum. The glaciers were in their prime then, deep and still moving fast— for glaciers. Then the climate changed, and the ice began to melt so rapidly that all the ice that would melt had melted within about 3,000 years. Glaciers move more slowly as they get thinner, so the glacier that dammed the Clark Fork River took longer to establish each successive ice dam, each successive lake. So it seems reasonable to suppose that the intervals between fillings of the lake were much shorter than average in the early years, then lengthened greatly towards the end as the glacier began to stagnate.

Pleistocene time, the last 2 million or so years, was the period of the great ice ages. Geologists believed for more than a century that Pleistocene time brought four major ice ages with three much longer interglacial episodes sandwiched between them. But more recent evidence, mainly from deep-sea cores, leaves geologists wondering how many ice ages really happened during Pleistocene time. At least eight according to some eminent authorities; as many as twenty according to others equally eminent. Our window into the Pleistocene past has become very murky indeed.

Whatever the actual number of major Pleistocene ice ages, the glaciated landscapes, including those of the Northwest, consistently show clear evidence of only the last two. The earlier ice ages left no record that survives in the modern landscape. Erosion constantly changes the landscape, so it is not at all surprising to find that it has erased the record of the older ice ages.

Radiocarbon dates depend upon the radioactive decay of carbon-14. They give good results, provided you can find samples of organic matter such as wood, charcoal, or bone that have not been contaminated with other carbon since they were deposited. Unfortunately, that has proved extremely difficult in the deposits left by Glacial Lake Missoula. In any case, virtually all of the original carbon-14 decays within 40,000 years, so the method does not work on material older than that.

Several quite different methods provide us with age dates on volcanic ash that are valid far into the past. Thin layers of volcanic ash have been found sandwiched within the Glacial Lake Missoula flood deposits of eastern Washington. Precise correlation of their chemical compositions with those of ash deposits of known age elsewhere in the region provide their ages. They bracket the time of Glacial Lake Missoula and its great floods between approximately 15,000 and 13,000 years ago. Even though those dates were obtained indirectly, they fit so reasonably into the larger picture of the most recent ice age that they seem likely to be about right.

Bubbles of air trapped in deep ice cores drilled in Greenland show a strong increase in their carbon dioxide content about 15,000 years ago. Evidently the composition of the atmosphere changed then. That was also when a sudden change in the climate ended the advance of the glaciers of the most recent ice age and started a great melt. By 12,000 years ago the cover of land ice was down to its present level and sea level had risen to its present level. That is extremely fast melting, considering that the great glaciers were as much as 12,000 feet thick in some areas. Many geologists now suspect that a sudden release of methane from the methane ice compounds in the sediments on the continental shelves caused a global greenhouse effect that changed the climate. Methane is a greenhouse gas that oxidizes to carbon dioxide in the atmosphere.

Glaciers of the earlier ice age grew considerably larger than those of the most recent ice age. The dates of that earlier ice age are the subject of much dispute. A guess of about 100,000 years ago for the time of its maximum is probably wrong, but it has the great advantage of being a nice round number that is easy to remember

Deep soil developed on Glacial Lake Missoula sediment, Ninemile Valley, western Montana.

and seems to offend a minimum number of geologists. Some day we may actually know. The dates of still earlier ice ages are even less known. For our purposes they hardly matter.

Sediments of the most recent ice age, including those deposited in Glacial Lake Missoula, typically show very little soil development. It takes more than the 12,000 or so years since that ice age finally ended for the tediously slow processes of weathering to convert the upper levels of freshly deposited sediment to a deep soil. Sediments from earlier ice ages typically have deep soils developed on them, soils that extend beyond plow depth. Scattered exposures of glacial lake sediment with a thick mantle of soil suggest that versions of Glacial Lake Missoula almost certainly existed before the most recent ice age.

So ice dams impounded several dozen lakes during the most recent ice age and an unknown number during at least one earlier ice age, possibly during several. The story of these glacial lakes is certainly much longer, more complex, and much richer than anyone now knows. So far, we have seen only the youngest and most obvious pieces of the puzzle.

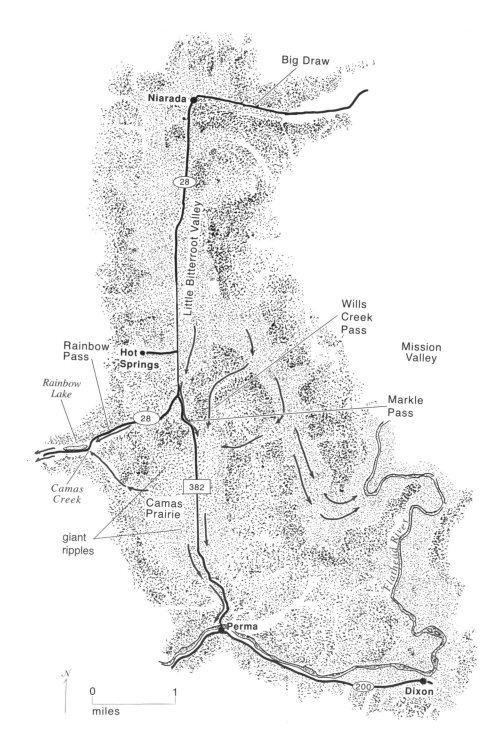

Big Draw

Niarada

28

Little Bitterroot Valley

Wills
Creek
Pass

Mission
Valley

Rainbow
Pass

Hot
Springs

Markle
Pass

*Rainbow
Lake*

28

*Camas
Creek*

382

Camas
Prairie

giant
ripples

Flathead River

N

Perma

0   1

miles

200

Dixon

*Camas Prairie and the Little Bitterroot Valley. Arrows indicate the
flow directions of the draining water.* —Adapted from Wallace Quadrangle,
1:250,000, U.S. Geological Survey, 1956

SETTING THE STAGE
FOR THE BIG SPECTACLE
The Little Bitterroot Valley

MOST GEOLOGISTS CONSIDER that the ridge south of Hot Springs, where Montana 382 crosses Markle Pass, defines the southern end of the Little Bitterroot Valley proper. So defined, the valley is about 3 miles wide and 15 miles long. But that is only part of the picture.

A southern extension of the Little Bitterroot Valley, about half as wide, reaches 15 miles farther south to the southwestern corner of the Mission Valley. Most geologists agree that it is an abandoned stream valley, but they consider it quite separate from the Little Bitterroot Valley, even though the two are perfectly continuous. The two valleys together look like about 30 miles of a single old stream valley big enough to hold a large river, a relic of a time when the climate was wet enough to maintain such a river.

The northern end of the Little Bitterroot Valley disappears under the Hog Heaven volcanic pile north of Niarada. Age dates on those rocks show that they erupted sometime around 35 million years ago, during Oligocene time. The old stream did not erode a valley through the volcanic rocks, evidently because it dried up before they erupted.

The old stream probably flowed during Eocene time, between about 60 and 40 million years ago. Eocene plant fossils and buried soils here and there in the Pacific Northwest leave no doubt that a warm climate prevailed during that time, a climate that was also wet enough to maintain large rivers. But the plant fossils, buried soils, and sedimentary rocks of more recent Oligocene time tell of a climate dry enough to stop the flow of rivers. Without flowing streams to carry them along, sediments accumulated in the floors of the dry river valleys.

Sediments deposited in Glacial Lake Missoula cover the floor of the Little Bitterroot Valley. They are deepest, between 200 and 300 feet, near Niarada at the north end of the valley. Montana 28 passes through patches of little hills eroded in pale silt, mostly glacial rock flour. No

The Little Bitterroot Valley and Camas Prairie, flooded
—Adapted from U.S. Forest Service (Jeff Silkwood), 1998

glaciers ever existed in the low mountains around the Little Bitterroot Valley, so where did that rock flour come from?

Montana 28 passes through the Big Draw just northeast of Niarada, crosses a divide, then continues east to Flathead Lake. Most geologists agree that the Big Draw looks like an old valley, much too narrow to be the continuation of the Little Bitterroot Valley. It looks like a tributary, an old canyon that emptied into the river that flowed through the Little Bitterroot Valley back in Eocene time. Its lower part is now filled with sediment, glacial outwash.

The drainage divide at the crest of the Big Draw is the top of a big glacial moraine. It was deposited at the edge of an ice lobe that crept west from the massive glacier that filled the Flathead Valley during the most recent ice age. Glacial meltwater swept sand, gravel, and rock flour that melted out of that ice down the Big Draw and into the Little Bitterroot Valley. Watch from Montana 28 for the meltwater channels on the gravel surface in the Big Draw— still clearly visible at least 12,000 years after the last glacial meltwater flowed that way. The fine rock flour continued to move south from the Big Draw, down the deeply flooded and southward-sloping floor of the Little Bitterroot Valley, probably as an undercurrent dense with suspended sediment. Some settled at the north end of the valley, near the mouth of the Big Draw, where it appears in the pale hills near Niarada. The rest spread across much of the valley floor, deep beneath the lake surface.

Enough rock flour spread across the lower south end of the Little Bitterroot Valley to make a continuous layer on the floor of Glacial Lake Missoula. Farmers there had been drilling flowing artesian wells for years when in 1916 O. E. Meinzer, one of the great pioneers of groundwater geology, showed that the impermeable cap of Glacial Lake Missoula rock flour maintains the high water pressure in the aquifer beneath the valley floor. The water is under pressure because it enters the aquifer through the gravel fill in the Big Draw, at the high north end of the valley. O. E. Meinzer was one of those who harshly criticized J Harlen Bretz in 1923, but he is much better and more fondly remembered for his work in groundwater geology.

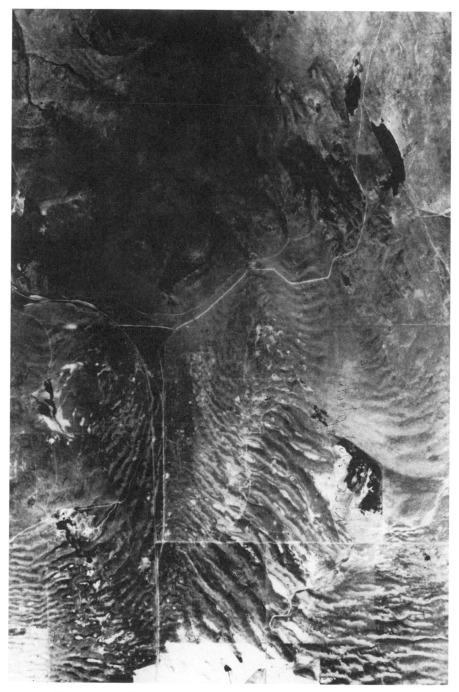

Vertical aerial photo of the giant ripples of Camas Prairie. J. T. Pardee first recognized them from such a picture. North is toward the top of the photo.

A QUESTION OF SCALE
The Giant Ripples of Camas Prairie

CAMAS PRAIRIE is a small basin immediately south and west of the Little Bitterroot Valley. A high ridge along Camas Prairie's northern side separates it from the southern end of the main Little Bitterroot Valley. Another ridge along the prairie's eastern side separates it from the southern continuation of the Little Bitterroot Valley and the Mission Valley. The forested ridge bounding the western side of Camas Prairie separates it from the Plains Valley, the Wild Horse Plains. Camas Prairie's southern end is wide open to the Flathead River.

The observant people who run cattle on Camas Prairie noticed long ago the hundreds of strange ridges as much as 35 feet high that curve across its surface like a wriggling herd of giant caterpillars. They wondered about them because they are not like anything on the floors of nearby valleys. They might still be wondering if J. T. Pardee had not explained those ridges in his paper of 1942.

Those ridges are the giant ripples of Camas Prairie, Pardee's premier exhibit of evidence that Glacial Lake Missoula drained catastrophically. Camas Prairie sports the most spectacular display of giant ripples associated with Glacial Lake Missoula, quite possibly the most spectacular giant ripples in the world. They have never been plowed, are easily accessible, and are easy to see—if you look at them with the help of J. T. Pardee.

When Glacial Lake Missoula emptied, water stored in the Little Bitterroot Valley poured south into Camas Prairie through Wills Creek

Pass, Markle Pass, and Duck Pond Pass, from east to west. A much lesser volume of water poured west from the Mission Valley into Camas Prairie through Big Creek Pass, on the eastern side of the prairie. Markle and Wills Creek Passes are deep notches in the crest of the ridge that separates the Little Bitterroot Valley to the north from Camas Prairie to the south. When Glacial Lake Missoula was filled to its highest shoreline, the water level was about 800 feet above the passes. The strong flow of water that poured through these passes into Camas Prairie scoured and plucked holes in the bedrock of both passes. Montana 28 passes a nearly circular pit about 50 feet deep in the crest of Markle Pass. Wills Creek Pass contains a group of ponds called the Schmitz Lakes. Duck Pond Pass, west of Markle Pass, is a broad swale that certainly carried a large volume of water across the ridge but was too wide to suffer much scouring.

After the water level in the Little Bitterroot Valley dropped below the level of the main passes, the remaining 400 feet of lake depth emptied through the broad avenue of the valley's southern continuation into the south end of the Mission Valley, thence into the valley of the Flathead River. The Little Bitterroot River, actually a little creek, now dribbles through the Little Bitterroot Valley and its·southern continuation.

Hydrologists recognize two main kinds of ripples in streambeds. Those that form under very fast water are called antidunes. They exist beneath the standing waves you often see in a flooding stream. The steepest slope of an antidune faces into the current. Under lesser flows, transition flows, the streambed is flat. Still slower water ruffles the bottom into much lower ripples called dunes. The steepest face of a dune faces downstream, away from the current. You never see antidunes in streambeds at times of low water because the stream erases them as it slows into transition flow.

Pardee included a section through a giant ripple in his paper of 1942. It looks like a section through an antidune. In 1976, James Lister studied the internal layering exposed by roadcuts and gravel pits in the high ripples below Markle and Wills Creek Passes. They are indeed antidunes, as their shape with the steepest slope facing into the current suggests. The lower ripples farther south are dunes. Transition flow evidently smoothed the flat area between them. Why did the antidunes survive? Probably because the flow across them stopped abruptly when the water level in the Little Bitterroot Valley fell below the passes.

Geologists were just beginning to make routine use of air photos when Pardee did his second study of Glacial Lake Missoula in the late 1930s. In the high vertical perspective of an aerial photo, the curving ridges on the floor of Camas Prairie look exactly like the little sand ripples in the bed of a stream—except for their outlandish scale. So Pardee recognized the giant ripples when he saw them in an aerial photo. Even so, I marvel at his perception. No one else in the whole world had even imagined such things as giant ripples in 1942. I marvel even more at his raw nerve in publishing a conclusion that he surely knew would leave many geologists wondering about both the giant ripples and his sanity for the next thirty years at least. According to V. R. Baker, geologists have since recognized more than one hundred trains of giant ripples on the floor of the old lake and along the paths of its floods to the Pacific. Everyone who admires them owes a debt of perception to J. T. Pardee.

Markle Pass and Wills Creek Pass. Stipple shows land above 3,400 feet. The arrows indicate directions of water flow. —Adapted from Markle Pass and Hot Springs Quadrangles, 1:24,000, U.S. Geological Survey

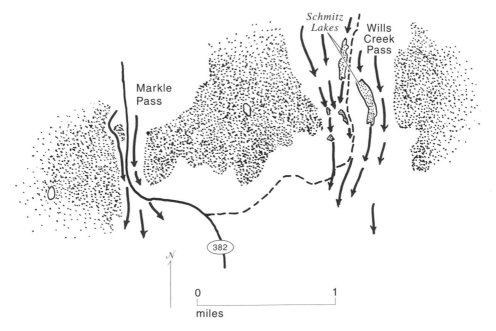

The best places to see a broad view of the giant ripples are along the narrow dirt road that connects Markle Pass and Wills Creek Pass, at the north end of Camas Prairie. An enormous debris fan with a nearly flat top and steep sides spreads below each pass. The trains of high ripples spread from the edge of each debris fan. The antidune ripples on the steep slope below the dirt road are as much as 35 feet high and 300 feet from crest to crest. They are composed of angular chunks of rock, some as big as shoeboxes. In the view from some places along that road, the antidune ripples look almost like high surf with the steep sides of the waves facing north. The high ripples end in the broad flat, leveled in transition flow, that contains a farm. Beyond that, a second train of much lower dune ripples trails off almost 9 miles into the southern distance.

Passes, debris dumps, and giant ripples in Camas Prairie —Adapted from Pardee, 1942

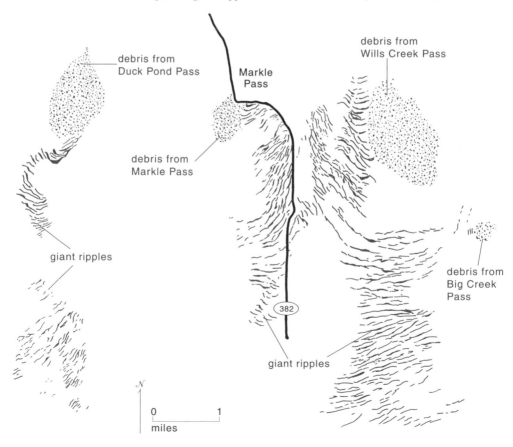

A MOST
EXTRAORDINARY BASIN
Rainbow Lake

THE PEOPLE WHO LIVE NEARBY call it Dog Lake; those who rely mainly on maps to know the country call it Rainbow Lake. By whatever name, this lake floods the largest bedrock basin within the former area of Glacial Lake Missoula. What made it?

Professors who teach introductory geology and the textbooks they use sternly instruct us that running water does not by itself erode bedrock but instead abrades the rock with the sediment it carries, like liquid sandpaper. That is why waterworn bedrock has a smooth surface shaped into flowing forms. But the drainage of Glacial Lake Missoula was not an ordinary stream flowing in the normal way. Rocks around Rainbow Lake show no sign of abrasion, nor do those in the scabland basins of eastern Washington. Instead, the strong currents appear to have plucked chunks out of the bedrock and swept them away. How did they do that?

Many geologists have wondered if the currents may have done it through cavitation. Watch a sheet of water flowing very fast over a fairly smooth surface, such as a large boulder or the concrete floor of the spillway below a dam. You are likely to see large bubbles appearing and disappearing along the surface between the water and the solid beneath. That is cavitation in action. Those bubbles do not contain air. They contain water vapor, and not much of that. For all practical purposes, they contain a vacuum. When the bubbles form, a vacuum exists above the bedrock beneath them. When the bubbles collapse, the water above them strikes the bedrock a heavy blow. The combination of the suction above the forming bubbles and the hammerlike blows beneath the collapsing bubbles can dismantle bedrock.

Cavitation worries engineers who design dam spillways, generator turbines, boat propellers, and many other structures exposed to rushing water. The question is whether it was effective in the case of Glacial Lake Missoula. Most hydrologists now think it unlikely that the currents in the draining lake and the Glacial Lake Missoula floods flowed quite fast enough to cause much cavitation, except perhaps very locally, and with minor effects. Kolks are the more likely culprits.

A kolk in full action

Dutch hydraulic engineers saw them first and called them colcs. The Germans spell it kolk, and so do we. However you spell it, they are extremely strong vertical vortices that develop within deep flows of very fast water.

Kolks are much fiercer than ordinary whirlpools or eddies in the same way that a tornado is much fiercer than a summer afternoon whirlwind. It is probably easiest to think of them as underwater tornadoes. And kolks treat big rocks in the same destructive manner that tornadoes bring to chicken coops and mobile homes. Dutch engineers have seen kolks hoist riprap blocks as heavy as automobiles and carry them off, twirling in their grip. Most hydrologists now think kolks plucked chunks out of the bedrock beneath the fastest flows as Glacial Lake Missoula drained.

When Glacial Lake Missoula drained, a strong flow of water poured west out of Camas Prairie along the line of Montana 28; another flow raced northwest up the valley of Camas Creek. They met just above Rainbow Lake. Those two flows of fast water likely were moving at different speeds, generating kolks as they sheared past each other. That is probably the simplest, and also the best, way to understand the plucking of the basin that holds Rainbow Lake, as well as many others in the region of Glacial Lake Missoula and along the path of its floods.

Rainbow Lake is no trifle. It is about 1.5 miles long, a fifth as wide, and as much as 35 feet deep, all within a bedrock basin. The small streams that enter it keep it full of water that stays fresh, though the lake has no surface outlet. It must drain by seepage through fractures in the bedrock. The bedrock is Belt mudstone, very hard to erode. Some of the plucked bedrock basins in the scablands of eastern Washington are considerably larger, but the comparison is unfair because they are in basalt, which plucks much more easily than Belt mudstone.

Montana 28 follows the path of the flow west of Rainbow Lake, across almost 3 miles of angular debris plucked from the lake basin and the swamps beside Montana 28. A straggly forest of stunted pine trees growing in the rubble field partially obscures the giant ripples that broadly corrugate the rocky surface. Watch for the gentle rise and fall of the ground on both sides of the road.

Three miles west of Rainbow Lake, the torrent of water with its load of debris ran directly into Locust Hill, a prominent landmark north of Montana 28. Locust Hill split the flow into two branches. The smaller branch went north into the Plains Valley, where it dumped a load of debris that made a sizeable bench with a very steep face. The larger branch went south into the valley of Boyer Creek, where it dumped a similar bench of coarse debris.

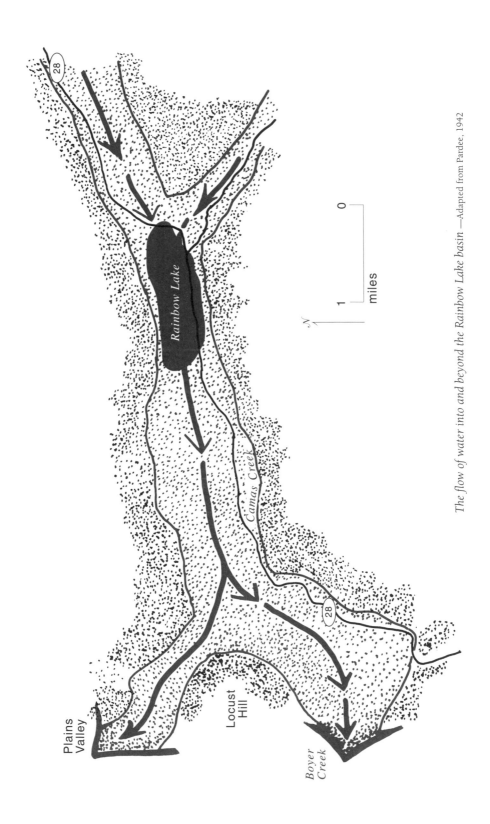

The flow of water into and beyond the Rainbow Lake basin —Adapted from Pardee, 1942

28

Rainbow Lake

Camas Creek

28

Plains Valley

Locust Hill

Boyer Creek

N

1   0

miles

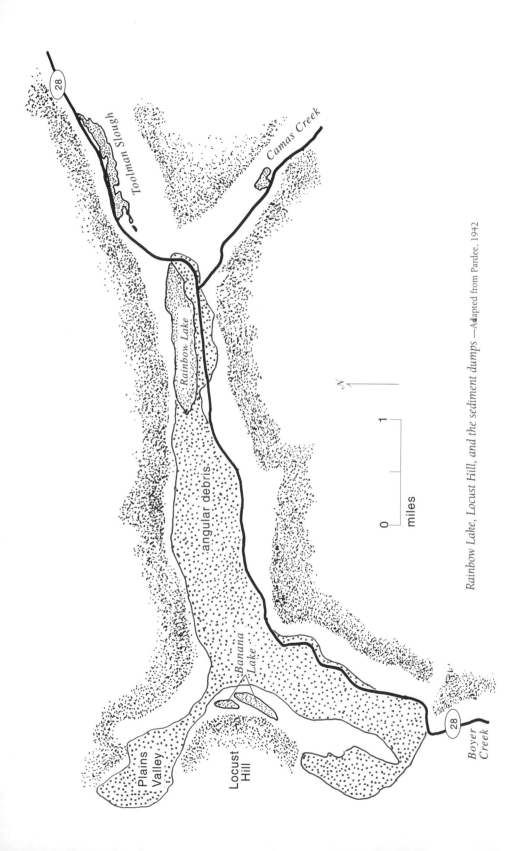

Rainbow Lake, Locust Hill, and the sediment dumps —Adapted from Pardee, 1942

*View from Montana 28: the steep western end of the sediment
dump that fills the valley of Boyer Creek, north of Montana 200*

Montana 28 follows the valley of Boyer Creek for about 3 miles
north of Montana 200. The valley looks perfectly normal until it
abruptly disappears beneath the steep face of the enormous dump of
debris that swept from Rainbow Lake and the passes above it. The
highway crosses the lower end of the bench in a long grade that passes
many excellent roadcuts. They expose a deposit of angular chunks in
crude layers that tilt down to the south. Those layers were deposited
on the front of the dump as it built its way down the valley of Boyer
Creek. All the rocks, large and small, are Belt rocks like those exposed at Rainbow Lake.

Most of the roadcuts also reveal large blocks of rock, still angular
and completely enclosed within the lesser debris. If the current had
rolled them, they would surely show far more rounding of their edges
and corners. Kolks must have carried them bodily and dropped them
where they now rest. Many of the deposits of coarse debris along the
path of the great floods contain such oversize angular blocks and
carry a litter of them on their surfaces.

Angular blocks within sediment exposed in a roadcut through the deposit in the Boyer Creek valley. The largest are about 5 feet long.

Banana Lake fills another bedrock basin just a few miles west of Rainbow Lake. Considered as scenery, Banana Lake offers nothing to inspire an artist. But it does flood a bedrock basin that curves gracefully around the east side of Locust Hill, and its banana-like shape is obvious on a map. Locust Hill stood against the flow like a boulder in a stream, while the strong current rushing around its upstream base eroded the channel that now holds Banana Lake. Streams commonly scour around the upstream sides of rocks in their beds. Banana Lake is a gigantic example, so grossly oversized that it is hard to perceive without the help of a map or an aerial photo.

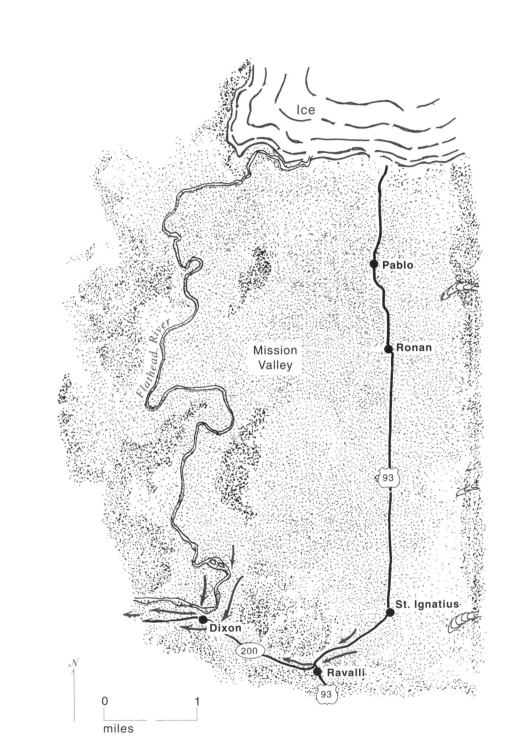

Ice

Pablo

Flathead River

Mission
Valley

Ronan

93

St. Ignatius

Dixon

200

Ravalli

93

N

0   1

miles

The Mission Valley. Arrows indicate direction of drainage flow.
—Adapted from Wallace Quadrangle, 1:250,000, U.S. Geological Survey, 1956

10

ICE, WATER, AND A RIVER
The Mission Valley, the Jocko Valley, and the Flathead River

AT THE HEIGHT of the most recent ice age, glaciers that came south from British Columbia completely filled the Flathead Valley north of Polson with ice. That ice was more than 4,000 feet deep near the Canadian border and dwindled southward to a thin edge just south of Polson. The terminal moraine of that glacier, the Polson moraine, makes a low ridge that crosses the valley from west to east just south of Polson. That moraine separates the Flathead Valley from the Mission Valley.

The Mission Valley is a broad basin west of the high and spectacular Mission Mountains, with their glacially gnawed peaks and deeply gouged valleys. The modest Salish Mountains define the valley's western edge and thinly separate it from the Little Bitterroot Valley. The straight eastern edge of the valley follows a fault along the base of the Mission Mountains. To the south, the Mission Valley terminates against the straight northern edge of a range of high hills, probably the line of a fault.

The high southern end of the Mission Mountains supported large valley glaciers during the most recent ice age. Large moraines deposited around their edges leave no doubt that the glaciers emerged from the mountains and onto the Mission Valley floor at an elevation below the upper level of most of the fillings of Glacial Lake Missoula. The lower ends of those glaciers certainly floated and snapped off in

49

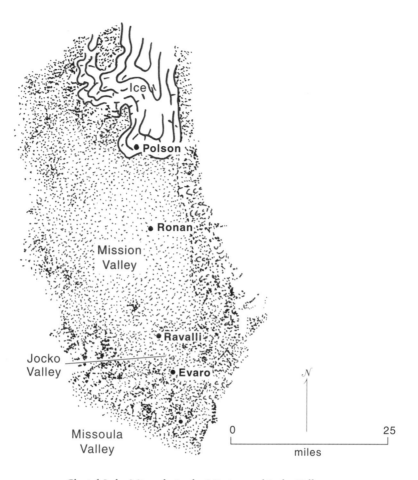

Glacial Lake Missoula in the Mission and Jocko Valleys
—Adapted from U.S. Forest Service (Jeff Silkwood), 1998

big pieces when Glacial Lake Missoula flooded the valley. And it is probably safe to assume that big pieces of the glacier that came down the Flathead Valley to the Polson moraine also broke off as the ice floated in the lake. The Mission Valley was one of the few parts of Glacial Lake Missoula that could have sported many icebergs.

The groups of boulders that abundantly litter the floor of the Mission Valley probably mark places where icebergs ran aground, then melted slowly in place. Some of the isolated boulders probably melted from stranded icebergs; others probably dropped from floating icebergs that were melting as they drifted in the lake. All the boulders are more or less rounded, as they should be if they were once carried in glaciers. And most of them lie on top of the ground surface, not embedded in it. They lie on the valley floor and on the low hills around it. Regardless of the details of their past history, all those boulders are souvenirs of Glacial Lake Missoula. We should cherish them and protect them from the apparently universal human tendency to get rid of stray rocks.

When Glacial Lake Missoula drained, most of the water stored in the Mission Valley poured directly down the Flathead River where it leaves the southwestern corner of the valley. The rest poured due south, down the steep slope of Ravalli Hill to meet the Jocko River at Ravalli, then flowed west into the Flathead River. The flow down Ravalli Hill carved a series of bedrock basins. Most are gone now, destroyed in the construction of U.S. 93. The ponds west of the highway in the National Bison Range are the only significant remnants. People pull off the road to watch the animals gathered around these watering holes.

The Jocko Valley lies directly south of the Mission Valley but is not in any sense its southern continuation. It may be another segment of the ancient stream valley that makes the Little Bitterroot Valley, now offset about 8 miles east along a fault. Movements on that fault, or swarm of faults, may also have emplaced the wall of rock that separates the north end of the Jocko Valley from the south end of the Mission Valley.

The Jocko Valley is about 12 miles long and fairly straight, with a generally northwest trend. It is about 2 miles wide, comparable in width to the Little Bitterroot Valley. It must have been a spectacular canyon when it was eroded, probably during Eocene time, some 50 million years ago. Then it filled with sediment during the two long arid periods between about 38 and 2 million years ago. The climate again became wet enough to maintain stream flow when the great ice ages began about 2 million years ago. Now the Jocko River and its tributaries are busily clearing the sediment from the old valley.

The Jocko River enters the Jocko Valley from the east, through a valley considerably larger than any the small modern river could erode. It is easy to imagine that this was the valley of a tributary that flowed into the large river that eroded the original Jocko Valley during the wet years of Eocene time.

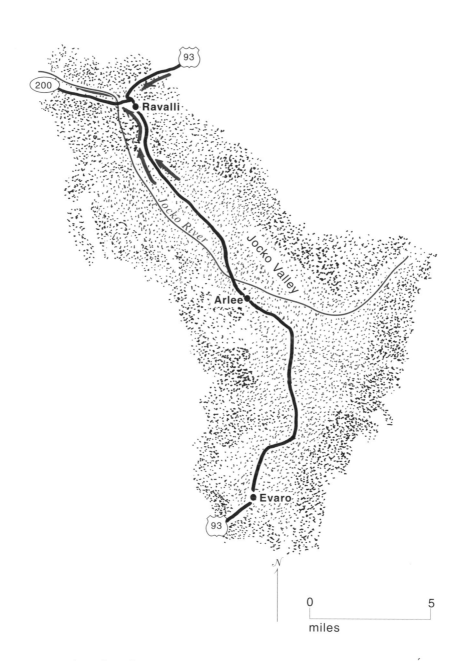

The Jocko Valley. Arrows show the directions of water flow as the lake drained. —Adapted from Wallace Quadrangle, 1:250,000, U.S. Geological Survey, 1956

U.S. 93 follows the old river valley through the Evaro area, a pass south of the Jocko Valley where sediment still fills the old valley. The pass is a broad expanse of flat land, often wet, but with no stream. The main hint of the filled canyon beneath comes from a few water wells that penetrate many hundreds of feet of the sediment fill. The filled valley appears to continue south to the Ninemile fault, which defines the straight north edge of the Missoula Valley.

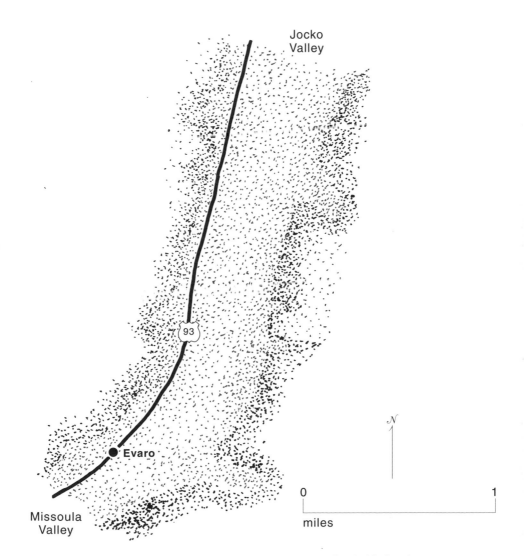

The broad pass at Evaro is probably an old canyon now deeply filled with sediment. —Adapted from Evaro Quadrangle, 1:24,000, U.S. Geological Survey, 1984

The broad benches of pale lake silt in the northern part of the Jocko Valley, east of the highway, consist largely of rock flour. Where did it come from? No glaciers scooped out the valleys or sharpened the peaks of the Jocko Hills east of the Jocko Valley. The Ninemile Mountains to the west held very little ice, and that only near the top of the ridge. The only large glaciers were in the Rattlesnake Mountains to the south, which separate the Jocko Valley from the Missoula Valley. Those glaciers may have spawned a few icebergs, but only when Glacial Lake Missoula filled to its higher levels. They certainly bled rock flour into the lake at all its levels. It must have moved in undercurrents heavy with suspended sediment that flowed north along the floor of the lake.

The only broad exit from the Jocko Valley is west down the Jocko River to the Flathead River just east of Dixon. That was the route most of the water followed when Glacial Lake Missoula drained. A much lesser volume of water may have flowed south along the line of U.S. 93 through the pass at Evaro, but perhaps not. The elevation there is about 3,950 feet, so water could have flowed through only when the lake was draining from its highest fillings, and then only in the earliest stage of drainage. The pass shows no sign of scouring, probably because it is so broad. I have seen no evidence that might tell whether the flow was south from the Jocko Valley into the Missoula Valley, or the other way.

Montana 200 follows the Jocko River west from Ravalli to its junction with the Flathead River just northeast of Dixon. Then it follows the Flathead River to its junction with the Clark Fork River just east of Paradise. Along most of that route, the Flathead River flows through a valley that looks too large to fit the modern river, probably another old valley eroded during the wet years of Eocene time. The river does not cross any exposed bedrock, and whitewater is conspicuously lacking, probably because it flows on soft sediments that filled the old valley during the millions of dry years that followed Eocene time.

Watch north of the Flathead River west of Dixon for the craggy lower valley walls with dark outcrops of bedrock. Strong currents that flowed while Glacial Lake Missoula was draining scrubbed the soil off those outcrops. An indistinct line separates the scrubbed lower valley walls from the unscrubbed higher slopes that still retain their original upholstery of soil. The same line exists on the south valley wall, but the trees make it hard to see. The line

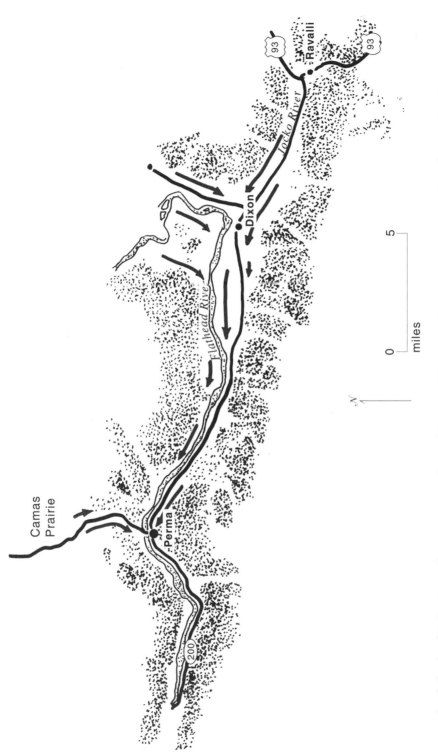

The Flathead and Jocko River valleys meet just northeast of Dixon, and the combined valley narrows. Arrows show direction of drainage flow.
—Adapted from Wallace Quadrangle, 1:250,000, U.S. Geological Survey, 1956

between scrubbed and unscrubbed valley walls is visible in many places but everywhere subtle.

Large deposits of sediment in the mouths of tributary valleys exist in many places along the Flathead and Clark Fork Rivers, but nowhere as conspicuously as along the north valley wall of the Flathead River between Paradise and Perma. J. T. Pardee called them high eddy deposits; the local people call them gulch fillings. They look almost like an earthen dam built across the mouth of the gulch, with a spillway breaching one side. Their surfaces are smooth and grassy. Pardee explained them as dumps of sediment that eddied into the slackwater in tributary gulches as the strong current of muddy floodwater thundered down the main channel. The gulch fillings between Paradise and Perma probably contain the soil the drainage currents eroded off the bedrock knobs west of Dixon.

Strong currents stripped most of the cover off these bedrock knobs west of Dixon and deposited sand and gravel on their downstream sides.

Gulch filling north of the Flathead River west of Perma

A large gulch filling exists in the mouth of a dry tributary south of Hellgate Canyon about halfway between Missoula and East Missoula. It is obvious enough, but its dense forest cover makes it harder to see than most of those along the Flathead River between Perma and Paradise.

The meaning of the upper surfaces of the gulch fillings is not entirely clear. They cannot record the upper surfaces of the drainage flows because many of them are well below the line between the scrubbed and unscrubbed valley walls. It seems more likely that they tell of immense loads of sediment in the lower levels of the drainage flows, less heavily burdened water in the upper levels.

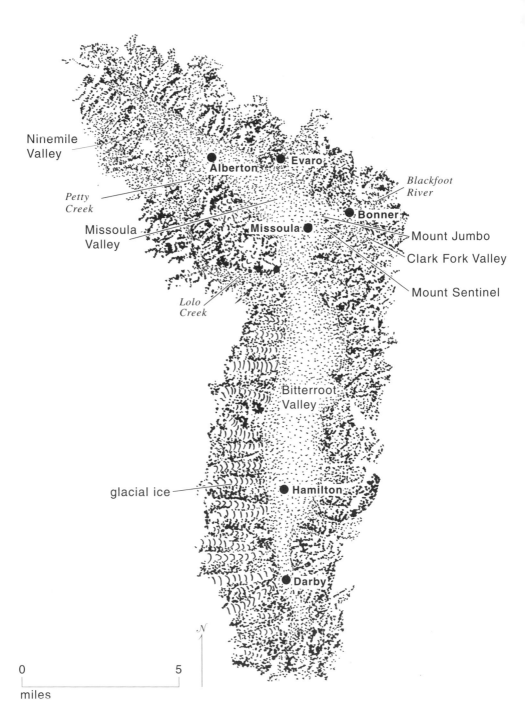

Ninemile
Valley

*Petty
Creek*

Missoula
Valley

Alberton

Evaro

*Blackfoot
River*

Bonner

Missoula

Mount Jumbo

Clark Fork Valley

Mount Sentinel

*Lolo
Creek*

Bitterroot
Valley

glacial ice

Hamilton

Darby

N

0   5

miles

Glacial Lake Missoula in the Bitterroot and Missoula Valleys
—Adapted from U.S. Forest Service (Jeff Silkwood), 1998

11

THE UNSCATHED VALLEYS
The Bitterroot and Missoula Valleys

THE BITTERROOT AND MISSOULA VALLEYS are broadly connected basins that opened during the same movements of the the earth's crust that raised the surrounding mountains. Geologists differ on exactly when that happened but agree it was sometime between 80 and 50 million years ago. Like other large valleys in western Montana, these filled with sediment during two long periods of desert climate, between about 38 and 17 million years ago and between about 15 and 2 million years ago. Sediments accumulate in desert basins because no stream exists to carry them out.

The Bitterroot and Clark Fork Rivers began flowing about 2 million years ago. That was the beginning of Pleistocene time, when the great ice ages began and the climate became much wetter. Those rivers have since removed as much as 800 feet of sediment from the valley floors, but at least 3,000 feet remain in some areas. The landscape of Eocene time, before the valleys filled so deeply with sediment, was far more spectacular than what we see today.

Along the west side of the Bitterroot Valley, the valleys of the high Bitterroot Mountains filled with glacial ice during the most recent ice age. Those in the much lower Sapphire Mountains along the east side of the valley did not. That explains why the Bitterroot Mountains are so much craggier than the Sapphire Mountains. The

Bitterroot Mountains still catch enough clouds to make their own weather, but the Sapphire Mountains probably never did.

Glaciers descended the valleys of the Bitterroot Mountains to levels as low as about 4,000 feet. Their lower ends could have floated when Glacial Lake Missoula was at its higher fillings, and ice may have broken off to make a few icebergs. But in most of the canyons north of Hamilton the lowest glacial moraines, which record the farthest reach of the glaciers, are a mile or more above the canyon mouths. So the long stretches of narrow stream canyon below the ice probably trapped any icebergs and prevented their drifting out into the broad expanse of the lake.

From Hamilton south, the floor of the Bitterroot Valley is high enough that glaciers emerged from the mountains and onto the broad valley floor, where they left large moraines studded with rounded boulders of granite. The ends of those glaciers may have spawned a few icebergs, but only when the lake reached its highest fillings. I have never seen stray boulders of granite on the broad benches along the east side of the Bitterroot Valley. Their absence suggests that no icebergs, which would have contained such boulders, made it across the valley, even though the east benches are on the downwind side of the valley.

When Glacial Lake Missoula drained, the Bitterroot Valley emptied north into the Missoula Valley, which drained west down the Clark Fork River. The entire Bitterroot Valley, including its north end, is wide open so the drainage currents left no obvious scars. Except for faint shorelines on some of the hillsides, the Bitterroot Valley looks as it would if Glacial Lake Missoula had never existed.

The Missoula Valley trends generally from east to west, at nearly a right angle across the northern end of the Bitterroot Valley. The Ninemile Valley is the Missoula Valley's western extension. Alpine landscapes that tell of ice age glaciation exist in the Rattlesnake Mountains on the northeast side of the Missoula Valley and in the northern end of the Bitterroot Mountains. Very small glaciers existed near the crest of the Ninemile Mountains north of the Ninemile Valley, mainly on the north side of the divide. No glaciers inhabited the other mountains around the Missoula Valley or lay on the valley floor. The lowest glacial moraines in the valleys of the Rattlesnake and northern Bitterroot Mountains are at an elevation of a little more than 4,000 feet, near or above the highest levels of Glacial Lake Missoula. That makes it highly unlikely that they could have spawned any icebergs.

Thin deposits of Glacial Lake Missoula sediments exist in many parts of the Missoula and Bitterroot Valleys, even in places remote from any mountain glaciers. Large deposits are conspicuous in the western part of the Missoula Valley, especially near the airport, which is on lake sediment. The rock flour around the airport likely originated in the glaciated Rattlesnake Mountains and traveled to the lake in glacial meltwater. It moved along the floor of the lake in currents heavy with suspended sediment, which ponded in this broad area of low ground.

When Glacial Lake Missoula drained, large volumes of water that had filled the Clark Fork drainage east of the Missoula Valley poured into the eastern end of the Missoula Valley through Hellgate Canyon and the lower Rattlesnake Valley. Those torrents left an abundance of souvenirs in the eastern end of the Missoula Valley. Meanwhile, water poured out of the west end of the valley through the Alberton Narrows, deeply scouring it. But the broad valley between shows almost no trace of the strong drainage currents. If it were not for the shorelines on the barren slopes along the northern and eastern sides of the valley and the patches of lake sediment on its floor, we would hardly know Glacial Lake Missoula ever filled the Missoula Valley.

Glacial Lake Missoula flooded the valley of the Clark Fork River east of the Missoula Valley to Garrison Junction, almost to the Deer Lodge Valley. None of the mountains along that way supported glaciers, but deposits of glacial lake sediments do exist here and there along the valley floor. The rock flour must have come from the Flint Creek Mountains, west of the Deer Lodge Valley, which did support large glaciers. Glacial meltwater could have carried the rock flour down the upper Clark Fork River from the east side of the mountains or down Flint Creek from their western slope.

Glacial Lake Missoula also reached north through the drainage of the Blackfoot River, all the way to Clearwater Junction, west of Ovando. Canadian ice came down the Swan Valley almost to Clearwater Junction, and Glacial Lake Missoula probably lapped onto it during its higher fillings. But the area of contact between glacier and lake was too small to spawn many icebergs. The immense glaciers in valleys of the Mission and Swan Mountains were high above the reach of Glacial Lake Missoula, but they surely supplied enormous amounts of

Glacial Lake Missoula sediment exposed in the north bank of the Clark Fork River, about 4 miles west of Drummond, beside Interstate 90

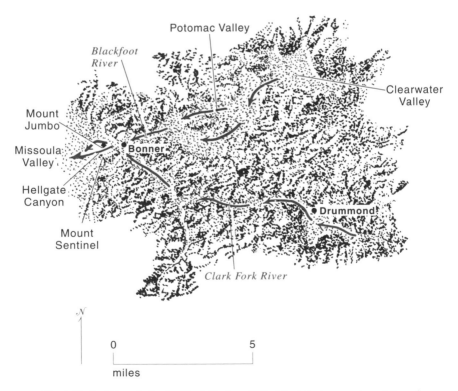

Glacial Lake Missoula in the Clark Fork and Blackfoot drainages above Missoula
—Adapted from U.S. Forest Service (Jeff Silkwood), 1998

meltwater and rock flour to the lake. The large expanses of ice that spread across the floor of the Blackfoot Valley at Ovando and places upstream were also beyond the reach of the lake, but they certainly bled rock flour into the Blackfoot drainage.

The valley of the Clark Fork River above the Missoula Valley stored an enormous volume of water when Glacial Lake Missoula existed. So did the valley of its tributary, the Blackfoot River, in the stretch between Bonner and Clearwater Junction, especially in the Potomac and Clearwater Valleys. All that water was destined to pour into the Missoula Valley when Glacial Lake Missoula emptied. A large proportion of it entered through Hellgate Canyon.

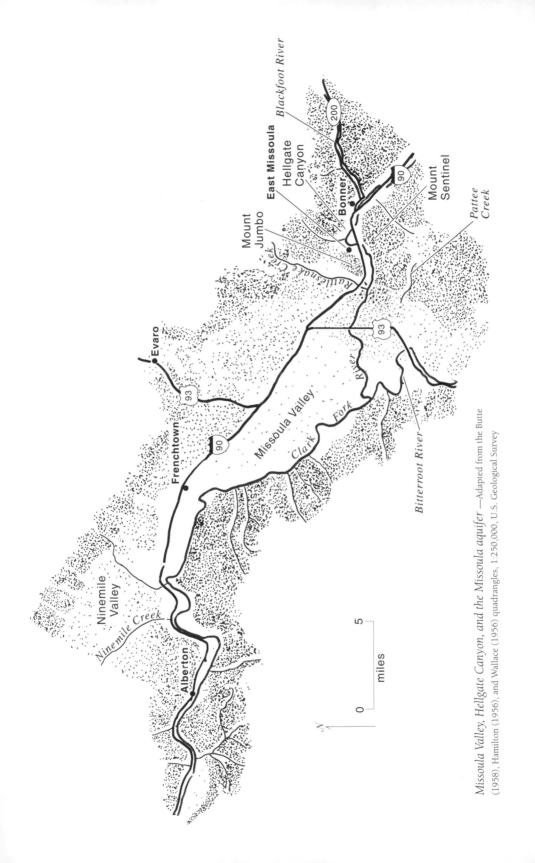

Missoula Valley, Hellgate Canyon, and the Missoula aquifer —Adapted from the Butte (1958), Hamilton (1956), and Wallace (1956) quadrangles, 1:250,000, U.S. Geological Survey

Blackfoot River

East Missoula

Hellgate Canyon

Mount Sentinel

Bonner

200

90

Pattee Creek

Mount Jumbo

Rattlesnake Creek

Evaro

93

Frenchtown

90

Missoula Valley

Clark Fork River

93

Bitterroot River

Ninemile Valley

Ninemile Creek

Alberton

0   5

miles

N

12

THE TIGHTEST SQUEEZE
Hellgate Canyon and the
Saddle North of Mount Jumbo

MISSOULA'S SOUVENIRS of Glacial Lake Missoula concentrate at the eastern end of the Missoula Valley, where water from the upper Clark Fork and Blackfoot drainages rushed through the narrow gorge of Hellgate Canyon and into the Missoula Valley. Water also came over the notch north of Mount Jumbo and into the valley of Rattlesnake Creek, thence into the Missoula Valley.

A rounded hill of bedrock just north of Hellgate Canyon, Mount Jumbo was named for a popular circus elephant of the 1880s but actually looks more like a caricature of a baked potato. The north end of Mount Jumbo descends into a low saddle eroded in very soft sediment. Like the soft valley sediment everywhere in western Montana, that beneath the saddle consists largely of volcanic ash. It is now mostly degraded into expandable clay, which weakens when it absorbs water and slumps easily.

The sediment in the saddle north of Mount Jumbo appears to fill a short segment of a long stream valley that a large river eroded, probably during Eocene time, about 50 million years ago. The sediment that later filled the stream valley was deposited between 40 and 17 million years ago. The much smaller Clark Fork River now follows that large river valley along most of its course between Drummond and St. Regis. Somehow, the short segment north of Mount Jumbo

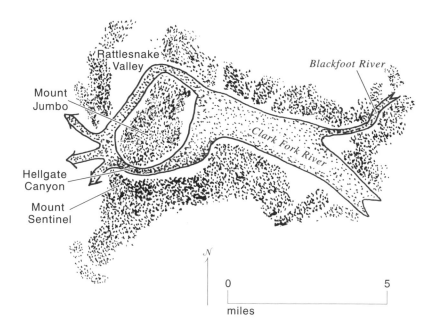

Drainage into the Missoula Valley through Hellgate Canyon and the saddle north of Mount Jumbo —Adapted from Bonner Quadrangle, 1:125,000, U.S. Geological Survey, 1903

remains filled. People driving west on Interstate 90 look directly ahead at that filled segment of the old valley as they approach East Missoula.

Why did the river that flowed through the old valley during the period of wet weather between 17 and 15 million years ago not follow that short segment of its course north of Mount Jumbo? Why did it erode Hellgate Canyon instead? Apparently, the top of the fill in the old valley was above the level of the stream that started flowing down the old valley 17 million years ago. A lower course was available along the path of Hellgate Canyon. Perhaps a large alluvial fan in the lower part of the Rattlesnake Valley raised the level of the fill and diverted the later river from the course of its predecessor. If so, Rattlesnake Creek and the drainage currents in Glacial Lake Missoula have since eroded the evidence. We can never know for sure.

When sediment filled the saddle to its original elevation of about 4,000 feet, and the lake was draining from its highest shoreline, water at least 200 feet deep flowed across the saddle. That water poured into the valley of Rattlesnake Creek, thence south the short distance to the Missoula Valley. Its flow eroded some of the soft valley sediment out of the saddle, reducing its elevation and opening the way for more water the next time the lake drained from a lower elevation. One after another, the long series of flows across the saddle reduced it to its present elevation of approximately 3,700 feet.

People who hike into the low saddle north of Mount Jumbo find two undrained depressions, each covering several acres. One is marshy at all seasons, the other only during wet weather. The last drainage currents over the saddle probably shaped those depressions into their present form. The two depressions ensure that all the water that falls on the saddle soaks into the expandable clay beneath. That explains why the slopes below both sides of the saddle are full of landslides and support so many trees.

Hellgate Canyon was certainly the tightest bottleneck any of the lake water passed on its way to the Pacific Ocean. It made a hydraulic dam that kept the water level east of the canyon much higher than in the Missoula Valley to the west. That steep gradient drove a powerfully erosive flow through Hellgate Canyon. Meanwhile, the bottleneck made the water level in the Blackfoot and upper Clark Fork drainages drop slowly, so the current through those valleys was less erosive than that pouring through Hellgate Canyon. That probably explains why the canyon walls of the Blackfoot and upper Clark Fork Rivers seem less dramatically scoured and plucked than we might expect at the junction of two large drainages.

Hellgate Canyon curves gently north, and the drainage currents hugged the outside of the bend, the south wall of the canyon. The floodwater scrubbed the bedrock on the south wall bare of soil and cut the extremely steep slope on that side. Its only cover is talus that accumulated since the lake last drained, sometime around 13,000 years ago. The north canyon wall, the south slope of Mount Jumbo, escaped that devastating current, and so retains most of its soil and a nice display of lake shorelines.

Montana 200 runs through East Missoula on a nearly level surface that rises to a low ridge at its eastern end, then drops abruptly down a very steep eastern slope, Brickyard Hill. The roadcut through the summit of the ridge reveals that the ridge is glacial rock flour, pale sediment so soft that you can dig it with a shovel, much too friable to withstand the vigorous drainage currents. That ridge makes it hard to interpret the flat surface to its west as either a river terrace or a lake bed deposit.

The ridge and the flat surface to the west are just where the broad valley of the Clark Fork River narrows as it enters the tight constriction of Hellgate Canyon. Something happened there, but what exactly? And why is this deposit rock flour instead of gravel? Perhaps it was laid down during the waning stages of the last drainage of the lake, when the current was too slow to carry gravel and finally dumped even its rock flour.

Mount Jumbo and Mount Sentinel, at the east side of the Missoula Valley, flank Hellgate Canyon to its north and south. The Glacial Lake Missoula shorelines etched on their grassy western slopes are so faint that on most days they are hard to see. But they boldly stand out as perfectly horizontal brown and white stripes when the snow is melting in late winter. And the early sun raking down the slopes on summer mornings raises them in strong shadow relief. Though this is the most famous display of Glacial Lake Missoula shorelines, countless other places along the Clark Fork drainage preserve shorelines as conspicuous as these.

It seems impossible for two people to count the same number of shorelines, or for one person to count the same number on different days or even at different times on the same day. The fainter shorelines seem to appear and disappear as the snow melts and the light changes. Despite all the uncertainties, attempts to count shorelines almost invariably leave the counter impressed with their number.

Could we be seeing extremely faint traces of old shorelines that date from ice ages before the most recent ice age? That idea is tempting, but probably wrong. Soil always creeps slowly down the slope. If the shorelines dated from different ice ages, we would expect the creeping soil to erase the older ones from the more active slopes. But the shorelines seem just as numerous on the weak clay slopes north of the Missoula Valley as on extremely stable slopes of Belt rocks. It seems likely then that they all date from the most recent ice age. How did the shorelines form, and at what season?

Glacial Lake Missoula shorelines on Mount Sentinel on one of their better days. Mudflows probably left the long scars that cross them. —D. W. Hyndman photo

The lake level probably rose too rapidly during the rapid summer snowmelt to allow waves time to carve a shoreline. The water level probably stabilized when the snowmelt stopped in the fall, and then storm waves could make a shoreline. But as the weather cooled into winter, the lake acquired a heavy burden of floating pack ice, which would still the waves. Perhaps most of the shorelines record storms that raised a heavy surf for a few days during the fall, after the summer snowmelt had stopped but before ice had spread from its shores. If the level of the next lake never rose to that of the lake before, waves could never erase the older shorelines. That may help explain why we see so many shorelines.

All the shorelines are very faint. Nothing suggests the violent crash of breaking waves. Trenches dug through Glacial Lake Missoula shorelines reveal precious little that looks like what we see in a trench dug through a modern beach. The lake shorelines show hardly a trace of wave deposition, hardly a pebble that the shuffling waves worked to a flattened shape. Nor do the old lakeshores include such common features as cliffs, sandspits, or deltas.

Faint as they are, all those shorelines on all those slopes assure us that before Glacial Lake Missoula existed the hills looked almost exactly as we see them now. The shorelines emphatically deny that Glacial Lake Missoula had much effect on the landscape, except very locally, in narrow canyons where the flow from the draining lake was exceptionally fast and turbulent. And all those surviving shorelines also assure us that the rate of erosion on the slopes that bear them is extremely slow.

At first glance, the long scars that track straight down the extremely steep west faces of Mount Sentinel and Mount Jumbo look like gullies. But water never runs down them. They have no rounded gravel in their floors. No tributaries join them. Several have no watershed that could conceivably supply surface runoff. And they do not extend above the highest shorelines. Similar scars exist on many other steep slopes that were once submerged in the lake. They look like mudflow scars. Much smaller but otherwise similar mudflow scars streak the slopes below old irrigation ditches that leaked enough to saturate the slope beneath them.

After Glacial Lake Missoula suddenly drained, those steep slopes were left saturated with water but without the buoyant support of the lake. They appear to have collapsed in long streaks, dumping their debris in small fans at the base of the slope. Some of the fans below Mount Sentinel are gone now, sacrificed to the road around the east side of the university campus. But the one east of the university married student housing survives almost intact, if you discount the effects of small children and their dogs digging over more than forty years.

Big blocks of Belt rocks, most of them angular and deeply embedded in the soil, litter the floor of the Missoula Valley within about a half mile of the mouth of Hellgate Canyon. Many weigh several tons. They are especially conspicuous on the campus of the University of Montana. They jut from many yards in the neighborhoods west of the campus and in the lower end of the Rattlesnake Valley. They rest on the soft sediments that fill the floor of the valley, so something carried them to where we now see them. To judge from the rock types, most came from the south wall of Hellgate Canyon, a few from farther east, perhaps as far as Bonner.

The angularity of those erratic blocks shows that something picked them up and carried them bodily. If the drainage currents had rolled them, that would have chipped off their sharp edges and corners, reducing them to rounded boulders. Many geologists long argued that those angular blocks were frozen into icebergs, which carried them along as they drifted across the lake. This line of reasoning contended that the prevailing westerly winds drove the icebergs to the east side of the valley, where they dropped the blocks as the ice melted in the summer sun. Unfortunately, that nifty story does not explain the evidence.

If icebergs did indeed carry those angular blocks, they would surely have left numbers of them on the western slopes of Mount Jumbo and Mount Sentinel. None exist there. Occasional large excavations for new buildings near the east end of the Missoula Valley reveal that the angular blocks are distributed through a matrix of rounded cobbles

Angular block embedded in The Oval, on the campus of the University of Montana. The numerals are 8 inches high.

and large pebbles, like raisins in a bagel. So they are both on and beneath the valley floor.

The angular boulders and their matrix of rounded cobbles and pebbles are clearly two distinct populations of rocks with different origins and histories. So why are they mixed into a single deposit?

The rush of water through Hellgate Canyon and the valleys just to the east swept rounded gravel from the submerged floodplain deposits. Meanwhile, kolks plucked angular blocks from the south wall of Hellgate Canyon, which has the raggedly quarried appearance typical of a plucked rock surface. The torrent emerging from Hellgate Canyon dumped rounded gravel and angular blocks together as it slowed upon spreading into the open expanse of the Missoula Valley. Deep excavations display a chaotically mixed deposit that shows hardly any sorting into layers—streams generally deposit sediment in neat layers of nicely sorted sand and gravel. Much of the neighborhood west of the university campus stands on what appears to be a giant gravel bar, but the cover of houses makes that very hard to see. The coarse gravel extends much farther beyond the mouth of Hellgate Canyon than do the angular blocks.

I believe that the flood dump of coarse gravel and boulders on the floor of the Missoula Valley makes the Missoula aquifer, which supplies Missoula with all its water. It is hard to imagine another major aquifer with a similarly catastrophic origin. The grain size of the sediments that make the aquifer diminishes westward.

When the drainage currents scooped the coarse gravel out of the floors of the river valleys just above the Missoula Valley and dumped it into the Missoula aquifer, they surely left a big hole where the gravel came from. The Clark Fork River promptly filled that hole with water to make a lake that snaked east along the valley floor, perhaps as far as Bonner. But a lake in the run of the river would soon fill with sediment, probably after just a few spring floods. Missoula revels in the nickname Garden City, which refers to the truck gardens that operated in Hellgate Canyon until the early 1970s. The soil there is soft and fine, without the gravel and angular blocks that infest the university area. It was probably deposited as mud in that filled and vanished lake.

13

THE WATER ROUTE TO PARADISE
Down the Clark Fork River to the Flathead River

EXCEPT IN ITS PASSAGE through the Alberton Narrows, the Clark Fork River flows through a broad and remarkably straight valley from the west end of the Missoula Valley to St. Regis. Long stretches of that valley look too large for the modern river. A larger river probably eroded them during the wet millions of years of Eocene time. Near the end of Eocene time, the climate became too dry to maintain rivers. The valleys of the Eocene rivers, deprived of flowing water, filled with sediment.

Another generation of rivers flowed through those valleys during another wet period between about 17 and 15 million years ago. They eroded much of the deep fill of sediment that accumulated after the valleys were first eroded during Eocene time. Then the climate again dried, and sediment once more accumulated in the valleys until about 2 million years ago. That was when the great ice ages began, and the region again became wet enough to maintain flowing streams. The Clark Fork River still has not eroded down to the valley floor of Eocene time. Its valley still contains part of its deep fill of soft sediments, which explains why so little hard bedrock is exposed in the riverbed below Missoula. The soft valley fill sediments are not the kind of material that makes whitewater.

About 30 miles west of Missoula, the Clark Fork River exits the Missoula Valley through the narrow canyon of the Alberton Narrows. That stretch of the river flows across a mass of hard Belt rocks in

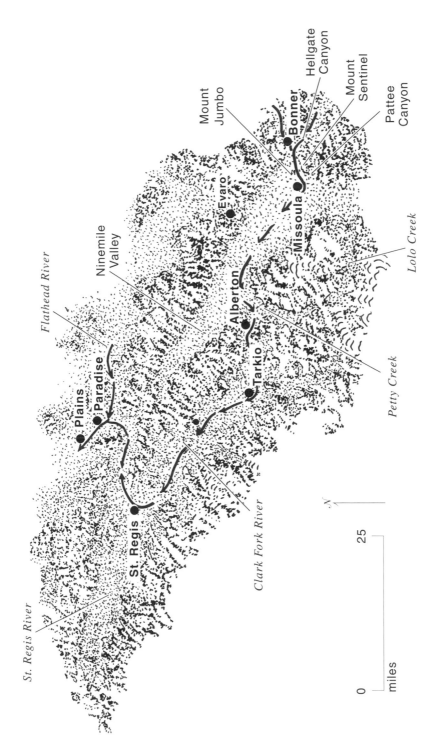

The flooded Clark Fork River between Missoula and Paradise —Adapted from U.S. Forest Service (Jeff Silkwood), 1998

Mount Jumbo

Hellgate Canyon

Mount Sentinel

Pattee Canyon

Bonner

Missoula

Evaro

Lolo Creek

Ninemile Valley

Flathead River

Alberton

Tarkio

Petty Creek

Paradise

Plains

Clark Fork River

St. Regis

St. Regis River

N

25

0

miles

foaming whitewater that rafters and kayakers greatly appreciate. They owe most of their fun to the Belt rocks, which would provide splendidly rough water even if Glacial Lake Missoula had never existed. But the torrent of the draining lake did make the whitewater a bit wilder. It stripped all the floodplain sediment off the big outcrops of red and green Belt mudstone in the valley floor near and just west of Alberton. Watch for them along Interstate 90. Also watch for the severely scrubbed valley walls that rise several hundred feet above river level.

The river exits from the Alberton Narrows into the far more spacious old valley in the area just downstream from the Cyr railroad tunnel at the lower end of the Alberton Narrows. The soft valley sediments deposited during the two long intervals of arid climate between about 40 and 2 million years ago are well exposed there, south of the river. The sediments clearly show that the river is again following an old valley that probably dates back at least to Eocene time.

The Clark Fork River exiting from the Alberton Narrows at Cyr. Old valley fill sediment is exposed in low hills immediately below the exit.

Watch immediately west of the Alberton Narrows for the large angular blocks that litter the wide and forested median between the lanes of Interstate 90. Kolks probably plucked these boulders from the Alberton Narrows, then dropped them on what appears to be a giant gravel bar. Giant bars commonly exist in open valleys immediately downstream from the mouth of a narrow canyon that once channeled an overwhelming flood. This one probably deflected the river into its course through the extremely narrow Alberton Gorge.

Large roadcuts north of Interstate 90 at the Tarkio Exit reveal a giant gravel bar deposited in the slack water downstream from Martel Mountain. Watch for coarse gravel in a wide range of sizes all mixed together. When the light is good, you can see crudely defined layers that slope steeply down to the west. They record former positions of the downstream surface of the growing gravel bar. Giant ripples cor-

The giant gravel bar at Tarkio —Adapted from Tarkio Quadrangle, 1:24,000, U.S. Geological Survey, 1983

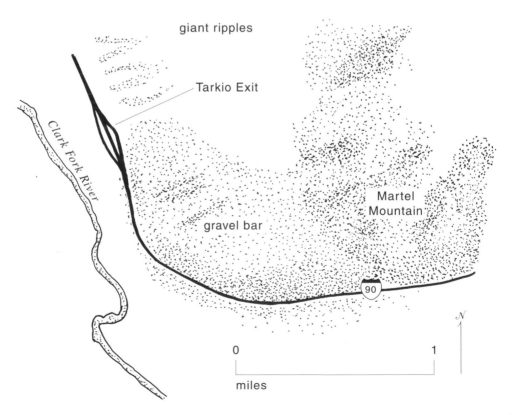

rugate the upper surface of the deposit, which is almost flat, and about 500 feet above the Clark Fork River. This is the only giant gravel bar in the valley exposed in a roadcut and so nicely visible from the highway.

Immediately west of the giant gravel bar at the Tarkio exit, giant ripples ruffle the surface of a river terrace. More such ripples exist on the river terraces downstream, especially between Tarkio and Lozeau. Watch for their gentle waves in the farm fields on both sides of Interstate 90. Just a few trees can make them very hard to see. It is nearly impossible to photograph the giant ripples, except from the air, because they are too big to fit into an ordinary picture.

The Clark Fork River follows the broad floor of the ancestral river valley on a northwest trend most of the way between the Alberton Gorge and St. Regis, where it makes a very tight right turn and heads northeast. Between St. Regis and its junction with the Flathead River, the Clark Fork River flows through a much narrower valley than that between Missoula and St. Regis. The river appears to have eroded this segment of its valley instead of inheriting it from a much older and larger river.

Gravel exposed in a roadcut at the Tarkio exit, westbound, from Interstate 90. The larger chunks are about the size of large baking potatoes.

Heavy forest cover obscures much of the evidence of the strong drainage currents along Montana 135 between Interstate 90 at St. Regis and Montana 200 just east of Paradise. Nevertheless, J. T. Pardee described areas of scrubbed bedrock valley walls and high deposits of coarse gravel. The scrubbed valley walls are easily visible from the highway, but they look a bit different from those elsewhere because this stretch of the river follows the layers of Belt rocks instead of cutting across them. Tree cover prevents easy viewing of the high deposits of coarse gravel, even though the highway crosses several of them.

The Flathead and Clark Fork Rivers join just east of Paradise. Both valley walls are thoroughly scrubbed for several miles downstream from their junction, the north valley wall most visibly so. The view north across the river from Montana 200 reveals a fantasy landscape of jagged outcrops cut from the bedrock. The original somber gray of the Belt rocks has become a reddish brown as the pyrite in them weathered to rusty iron oxide.

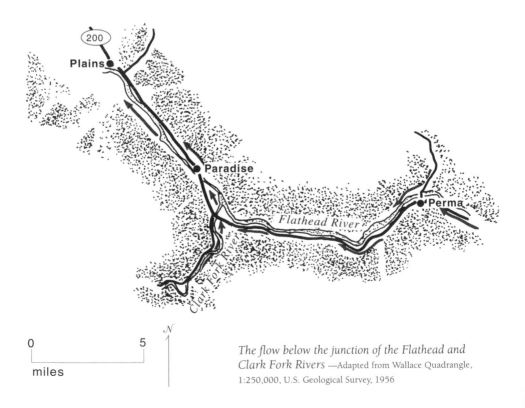

miles

0   5

The flow below the junction of the Flathead and Clark Fork Rivers —Adapted from Wallace Quadrangle, 1:250,000, U.S. Geological Survey, 1956

Watch especially for the remains of a new channel mostly hidden behind a screen of eroded rocks. The river would now follow that channel if the drainage currents had cut it a bit deeper. It was a shortcut across a big bend in the valley. Those currents cut similar short-cut channels across bends in several other places, but none of them became new channels.

Why did the drainage currents focus so much of their energy on this stretch of the valley walls?

We commonly see the waters from two different rivers flowing side by side before they finally mix into a single flow. In places where a clear stream joins a muddy stream, the combined flow is obviously clear on one side and muddy on the other. A close look generally reveals a trail of eddies along the line between the two flows, where the faster flow shears past the other. That is probably what happened where the drainage flows of the Flathead and Clark Fork Rivers met. But those two flows were so enormous and so fast that kolks formed, instead of mere eddies, along the shearing surface between them.

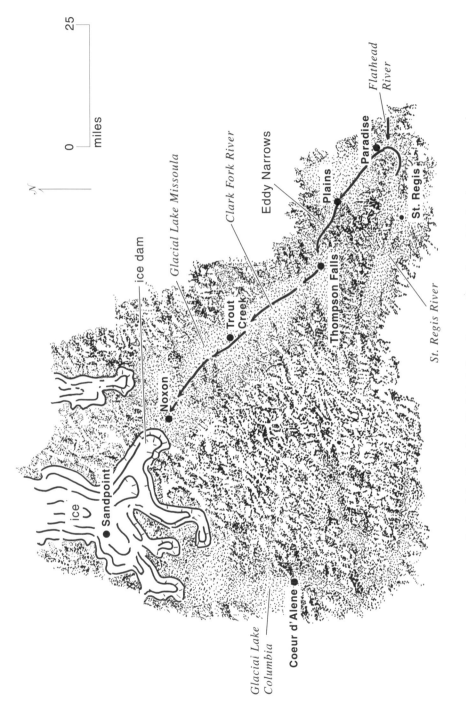

The flooded valley of the Clark Fork River between Paradise and the ice dam

14

TAKING THE MEASURE OF THE DRAINAGE CURRENTS
The Clark Fork River between Paradise and Sandpoint

THE VALLEY BELOW PARADISE carried all the water that drained from all the parts of Glacial Lake Missoula. Eddy Narrows, between Plains and Thompson Falls, is the best possible place to estimate that flow. The valley there is nearly straight for about 10 miles and nearly uniform in width. That is where Pardee took the measure of the drainage currents in Glacial Lake Missoula and of its humongous floods.

The discharge of a stream is the amount of water flowing through its channel. Stream hydrologists calculate discharge by multiplying the cross-sectional area of the channel by the speed of the flow. It is easy to measure the discharge of a small stream flowing through a clear channel. Measuring the discharge of a gigantic stream that briefly flowed many thousands of years ago is quite another matter. First, you must reconstruct the stream's channel and measure the area of its cross section. Then, you must estimate the speed of a current that no longer flows. J. T. Pardee did all that in the late 1930s, when he estimated the discharge through the Eddy Narrows as Glacial Lake Missoula emptied.

Pardee located the upper level of the drainage flow at the line between the scrubbed and unscrubbed valley walls. That rather fuzzy

81

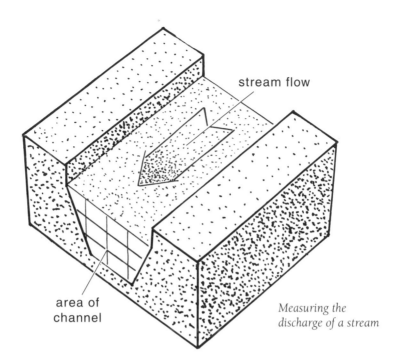

stream flow

area of
channel

*Measuring the
discharge of a stream*

boundary is about 1,000 feet above the valley floor, little more than half the maximum depth of the lake in that area. But the upper half of the lake water was spread out through the tributary valleys on either side of Eddy Narrows and above many of the mountains. That higher half had to drain into Eddy Narrows before it could find a clear channel. Until then, its downstream flow could not add much to the total.

Once Pardee had located a line between scrubbed and intact valley walls, he used ordinary topographic maps to draw profiles across the fast part of the drainage current. Those reconstructed sections of the channel had carried the current. That was fairly simple. Estimating the flow speed through that channel was much harder.

Stream hydrologists experiment with laboratory flumes to find the relations between the gradient of the streambed, the largest particles the flow can move down that slope, and the flow speed of the current. Pardee could measure the gradient of the bed directly, and could find the largest rocks the current moved in the deposits of coarse gravel on the valley floor just east of Thompson Falls. Then he and Walter Langbein, a distinguished hydrologist who also worked for the U.S. Geological Survey, used the experimental data to infer a flow

Scoured valley wall along the Eddy Narrows of the Clark Fork River

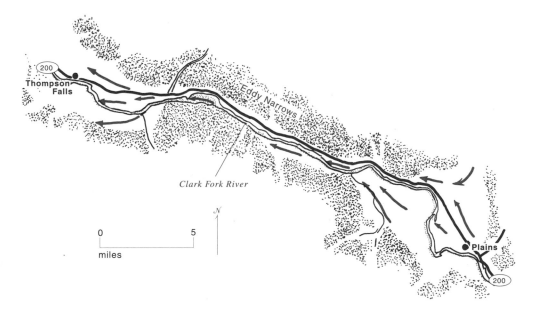

Eddy Narrows. Arrows indicate direction of drainage flow. —Adapted
from Wallace Quadrangle, 1:250,000, U.S. Geological Survey, 1956

speed of approximately 58 miles per hour. They multiplied that figure by the approximate sectional area of the channel to arrive at an
estimate of 9.46 cubic miles of water per hour for the maximum
drainage flow through the Eddy Narrows. Their jump from the scale
of a laboratory flume to that of Eddy Narrows was, to put it mildly,
what engineers call a "hairy extrapolation."

Hydrologists normally measure stream discharges in cubic feet or
cubic meters per second, certainly not in cubic miles of water per
hour. To put his fantastic number in some semblance of perspective,
Pardee noted that the maximum discharge ever measured on the Mis-
sissippi River was 0.05 cubic mile of water per hour. The average
discharge of the Amazon River, the world's largest, is about 0.014
cubic mile of water per hour. P. Weis and W. Newman of the U.S.
Geological Survey estimated in 1971 that the peak discharge through
Eddy Narrows was approximately ten times the combined flow of all
the rivers of the world. The drainages of Glacial Lake Missoula were
thumping great occasions.

In a raw way of thinking, a discharge between 8 and 10 miles of
water per hour could empty a lake of 500 cubic miles in about two
days, but of course that was not what happened. The discharge slowed
as the dropping water level diminished the cross-sectional area of the
channel. So it took more than two days to empty the lake, at least
several days, perhaps a week.

Pardee's estimate aroused at least as much skepticism as wonder,
even among geologists firmly convinced that enormous flows had
indeed passed that way. Various later attempts to measure the discharge through the same segment of the valley, most notably those by
V. R. Baker, broadly confirm Pardee's pioneering efforts. So far as anyone can tell, Pardee was at least approximately right, certainly well
within the ballpark.

The valley walls in Eddy Narrows are scoured to bare bedrock nearly
without any cloak of soil remaining. The same ferocious drainage
currents that scrubbed the valley walls also scooped the gravel out of
the floodplain and spread it across the broad floor of the valley just
east of Thompson Falls. Watch for the roadcuts in gravel along Montana 200. Each drainage probably left a long basin filled with water
in the floor of Eddy Narrows. Those lakes lasted only as long as it

took the river to fill them with fine sediments, probably just a few normal floods. That explains the stark contrast between the bare valley walls and the flat valley floor filled with fertile soil.

Large boulders like those that litter parts of the valley floor in Eddy Narrows do not appear in the big gravel deposits in the valley just east of Thompson Falls. That makes it seem likely that they simply tumbled down the high cliffs long after the lake drained, instead of travelling in the grip of kolks in the drainage currents.

The strong drainage currents left very few obvious scars along the broad valley below Thompson Falls. That long stretch of valley is mostly open, without the narrow chutes and large tributaries that

The ragged north valley wall and fertile valley floor along the Clark Fork River at Eddy Narrows

spawned swarms of kolks at so many places upstream. And the thick forest cover effectively conceals scoured bedrock, gulch fillings, and lake shorelines.

The river reaches the eastern end of the vanished ice dam just west of the border between Montana and Idaho. People traveling west on Montana 200 see high mountains on the northern horizon until they approach Lake Pend Oreille, where the skyline abruptly opens. That is the view north up the Purcell Trench, the route of the ice that once flowed south and dammed the Clark Fork River. Like most large geologic features, the trench and the former existence of great glaciers are very hard to perceive from the ground. We humans, puny creatures that we are, need the help of a map. Pardee had reasonably decent maps when he visualized the former ice dam back in 1910, but his scientific vision remains a wonder.

15

GATEWAY TO THE SCABLANDS
Rathdrum Prairie

ALTHOUGH STREAMS FOLLOW IT, no stream could have eroded the great trough of the Purcell Valley. Its straight furrow south through British Columbia and into the northern Idaho Panhandle must have something to do with the structure of the earth's crust, probably something fairly profound. The Purcell Valley dwindles south of Pend Oreille Lake into Rathdrum Prairie, which ends abruptly at the wall of hills west of Coeur d'Alene Lake, south of Interstate 90. U.S. 95 follows Rathdrum Prairie south from Sandpoint, past Pend Oreille Lake where the ice dam once lay, and along the path of the great floods down the Rathdrum Arm of Glacial Lake Columbia.

Glacial Lake Columbia was another large lake impounded behind an ice dam. It flooded a large area that included most of Rathdrum Prairie, all of the Spokane Valley, and the valley of the Spokane River west to the upper end of Grand Coulee. The elevations of its overflows into the Telford–Crab Creek scabland northwest of Spokane (chapter 18) show that the surface of Glacial Lake Columbia stood at an elevation of approximately 2,400 feet above sea level. That means it flooded Rathdrum Prairie to within a few miles of the ice dam that impounded Glacial Lake Missoula. Every time Glacial Lake Missoula floated and broke its ice dam, a wall of water rushed south down Rathdrum Prairie and plunged into Glacial Lake Columbia. In the earlier events, that wall of water was as much as 2,000 feet high and had as much as 500 cubic miles of water behind it. In later events, its

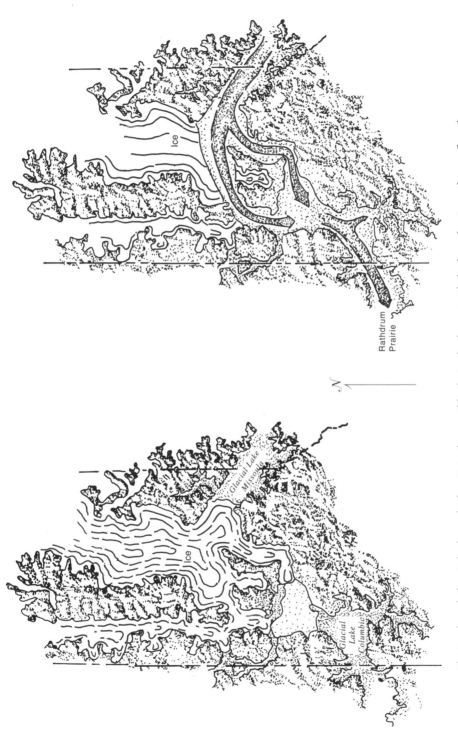

The ice dam, before it broke and after. It is also possible that only the east side broke, releasing only one flow down Rathdrum Prairie. —Adapted from U.S. Forest Service (Jeff Silkwood). 1:1,000,000, 1998

height and mass diminished as the ice dam thinned and floated at progressively lower levels of the lake.

Some geologists have suggested that Glacial Lake Missoula may have drained in a less than catastrophic manner as water poured over the top of the ice dam or gushed out from beneath. They have argued that no obvious evidence of flood scouring exists in Rathdrum Prairie, therefore no catastrophic floods ever flushed through it. It is hard to imagine that the lake could rise high enough to overflow the ice dam without floating it. It is especially difficult to imagine that the lake could drain as water leaked beneath the ice dam. The great depth of water behind the dam exerted enormous hydraulic pressure at the base of the ice. The moment the ice began to detach from the valley floor, the high pressure of the fingers of water injecting beneath would have popped it loose and burst the ice dam.

Good field evidence of catastrophic flows down Rathdrum Prairie does exist. A train of giant ripples on the gravel deposits east of Spirit Lake formed under strong currents that flowed through Glacial Lake Columbia as Glacial Lake Missoula dumped into it. Furthermore, the tangled web that the scabland flood channels and deposits weave across the map of eastern Washington shows that the floods did not tamely follow the established stream courses to the Columbia River. They sloshed right across the landscape. Water suddenly filled the broad valleys much faster than the streams could drain them. The valleys filled so fast that their overflows eroded deep channels across divides. The bottom definitely dropped out of the bathtub.

People who have watched big pieces of ice breaking off the ends of floating glaciers commonly compare the noise to that of heavy artillery fire. Imagine the shattering cannonade that must have accompanied the breakup of an ice dam at least 10 miles across. The water pouring through the breach in the ice dam was certainly brown with mud and full of great blocks of ice, broken trees, ruined mastodons, and assorted other debris. The ground surely trembled underfoot, and the racket must have been deafening for miles around. Without a doubt, these events were among the great spectacles of the most recent ice age.

Great torrents of glacial meltwater draining from the regional ice to the north certainly poured down Rathdrum Prairie long after the last Glacial Lake Missoula floods had passed. Those later torrents dumped untold millions of tons of gravel onto the prairie. Much of

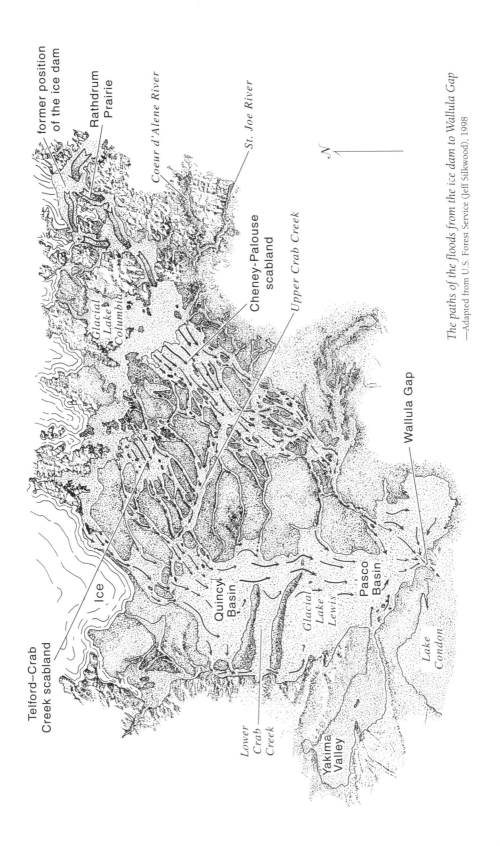

Telford–Crab Creek scabland

former position of the ice dam

Rathdrum Prairie

Coeur d'Alene River

St. Joe River

Cheney-Palouse scabland

Upper Crab Creek

Glacial Lake Columbia

Ice

Quincy Basin

Lower Crab Creek

Glacial Lake Lewis

Pasco Basin

Wallula Gap

Yakima Valley

Lake Condon

N

The paths of the floods from the ice dam to Wallula Gap
—Adapted from U.S. Forest Service (Jeff Silkwood), 1998

the gravel that covers the valley floor today undoubtedly belongs to those younger deposits, which show no evidence of flood scouring but probably bury such evidence.

Even though Rathdrum Prairie is not an eroded valley, a large river likely flowed through it long before ice age glaciers deranged the drainage. The valleys of tributary streams that flowed into that vanished river from the mountains on either side of Rathdrum Prairie are still perfectly obvious. If we can judge from the way they angle in from the north and east, the old river flowed south through Rathdrum Prairie, then west toward Spokane. The old valley continues west out of Rathdrum Prairie to the area of Post Falls, where it disappears beneath the much younger basalt lava flows of eastern Washington. I know of no direct evidence that might tell when that vanished river flowed, but suspect it was probably during Eocene time, between about 60 and 40 million years ago. The climate of our region was wet enough then to maintain large rivers.

The sudden rush of water from Glacial Lake Missoula dumped enough gravel into Rathdrum Prairie to block the mouths of the old tributary valleys. That gravel impounds Hayden, Hauser, Spirit, and Twin Lakes in the lower ends of those valleys. The lakes drain mainly by seepage through the coarse gravel.

The sequence of events that created Coeur d'Alene Lake began when Glacial Lake Columbia filled the Spokane Valley and backed the St. Joe and Coeur d'Alene Rivers much farther upstream than Coeur d'Alene Lake now does. The long series of Glacial Lake Missoula floods dumped a large deposit of gravel at the submerged lower end of Rathdrum Prairie. It buried the old channel of the St. Joe River.

Coeur d'Alene Lake as we know it first appeared when Glacial Lake Columbia finally drained through the Columbia River near the end of the most recent ice age, and the Spokane Valley emptied. That left Coeur d'Alene Lake impounded behind the dump of flood gravel at the lower end of Rathdrum Prairie. The new lake received the entire flow of the St. Joe and Coeur d'Alene Rivers, far more water than could drain by seepage through the gravel dam. So the lake overflowed, establishing a spillway across the gravel. Of course, the overflow followed the lowest available path, but that path was not above the buried former channel of the St. Joe River. So the overflow entrenched a course through the gravel, missed the buried river channel, and came down onto hard bedrock, which did not permit the

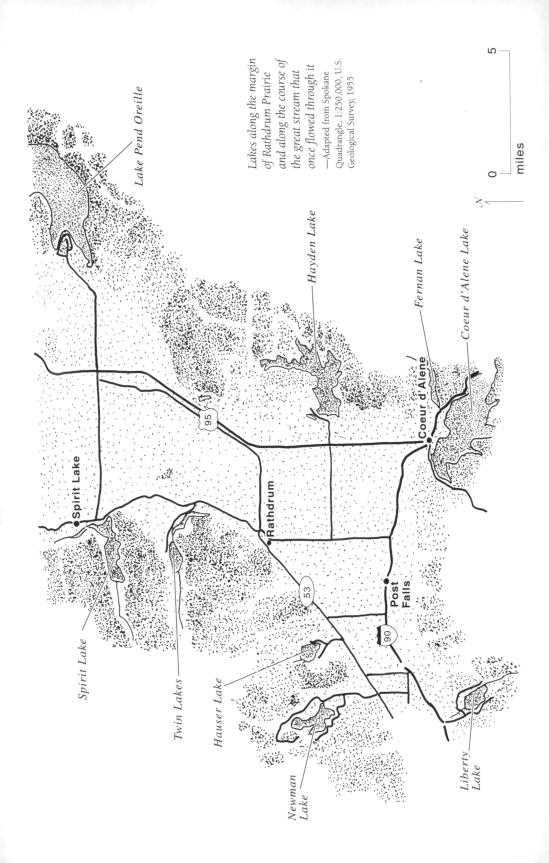

Lakes along the margin of Rathdrum Prairie and along the course of the great stream that once flowed through it
—Adapted from Spokane Quadrangle, 1:250,000, U.S. Geological Survey, 1955

miles

0   5

N

Lake Pend Oreille

Hayden Lake

Fernan Lake

Coeur d'Alene Lake

Coeur d'Alene

Spirit Lake

Rathdrum

95

53

Post Falls

90

Spirit Lake

Twin Lakes

Hauser Lake

Newman Lake

Liberty Lake

overflow to erode the spillway deep enough to drain the lake. The overflow spillway is now the Spokane River between Coeur d'Alene Lake and Post Falls. The present Spokane River west of Post Falls is the lower part of the old St. Joe River.

Liberty Lake, like the other lakes in this chain, is in a tributary of the old river and lies behind a dam of flood gravel dumped on the floor of Glacial Lake Columbia. It drains mainly by seepage through the permeable gravel. West of Liberty Lake, all traces of the old river vanish beneath basalt lava flows.

Columnar and hackly basalt in Moses Coulee

16

FLOODS OF BASALT
Lava Flows and Crustal Arches

BASALT, PLAIN BLACK VOLCANIC BASALT, is important to this story because the scablands would look quite different if they were eroded in any other kind of rock. The Glacial Lake Missoula floods crossed basalt lava flows from Rathdrum Prairie all the way to the Columbia Gorge. The only exception is a small area around the north end of Grand Coulee, where the floods cut through the basalt and into much older granite.

Basalt lava shrinks as it crystallizes into solid rock, then shrinks a bit more as the solid rock cools. That shrinkage typically opens a pattern of vertical fractures that make palisades of columns on cliff faces and in roadcuts. If the rock is black or brownish black and breaks into rows of more or less vertical columns, it is basalt. You can recognize it from a mile away.

Basalt columns range in diameter from fencepost size to barrel size. They may have four to seven sides, most commonly five. The columns generally extend all the way through lava flows less than about 50 feet thick, but most of the basalt flows in the Columbia Basin are at least 100 feet thick. These thick flows typically have a palisade of columns at the base, another at the top, with a zone of irregularly fractured, hackly basalt between them. The columns in the lower palisade are generally thicker and shorter than those above the hackly fracture zone.

Even though columnar basalt is a very hard rock, it is peculiarly vulnerable to flood erosion. The rushing water plucks the columns out of a cliff face and sweeps them away. The hackly basalt is more closely fractured, but its irregular pieces interlock in a more complex pattern that does not so readily lend itself to plucking. That makes it less vulnerable to flood erosion.

Most of the basalt in the Columbia Basin erupted as enormous lava flows, flood basalts, during middle Miocene time, between about 17.2 and 15.5 million years ago. Flood basalt eruptions commonly produce more than 100 cubic miles of lava, covering tens of thousands of square miles in just a week or so. When the great eruptions ended, the basalt surface of eastern Washington was nearly smooth and had the general form of an extremely shallow saucer, the Columbia Basin. Most geologists suspect that this basin assumed its form as the earth's crust sagged under the tremendous weight of the lava flows.

Sometime after the big eruptions ended, strong tectonic forces directed from the south began to shove great slices of the layered basalt flows north along rather gently dipping faults. The basalt above the

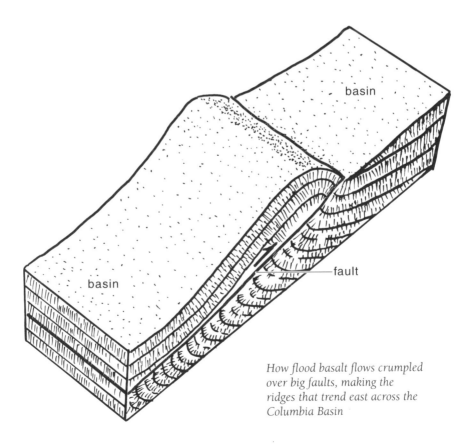

How flood basalt flows crumpled over big faults, making the ridges that trend east across the Columbia Basin

faults rumpled into long arches, called anticlines, with crests that trend generally east. Each anticline is now a long ridge, very steep on its north flank, gently sloping on its south flank. Broad basins that also trend generally east separate the sharp anticlinal ridges. The ridges are highest and sharpest in the west.

The ridges that most influenced the Glacial Lake Missoula floods were the Horse Heaven Hills in the south, the Saddle Mountains ridge to the north, and the Frenchman Hills ridge still farther north. The Horse Heaven Hills presented a long barrier, with the Wallula Gap its only pass. The Saddle Mountains ridge and the Frenchman Hills ridge are high in the western part of the flooded area but dwindle eastward. They separate the Pasco Basin in the south from the Quincy Basin farther north.

The Columbia and Yakima Rivers were already established in their present valleys when the faults began to raise the anticlinal arches into ridges. They were able to maintain their courses by eroding their channels as fast as the arches rose across their paths. The rivers now pass through those ridges in several narrow canyons, of which Wallula Gap through the Horse Heaven Hills near Pasco is the most familiar—and the most important to this story.

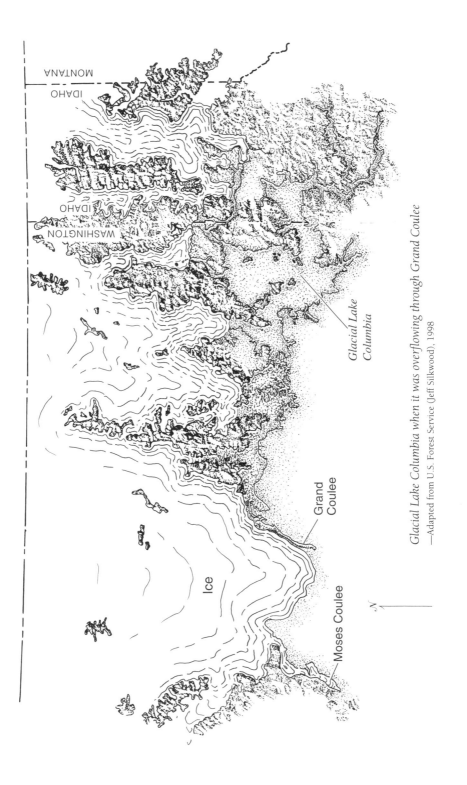

Glacial Lake Columbia when it was overflowing through Grand Coulee
—Adapted from U.S. Forest Service (Jeff Silkwood), 1998

17

MOSES COULEE
Okanogan Ice and Glacial Lake Columbia

MOSES COULEE is generally parallel to Grand Coulee, about the same length, and about 10 miles farther west. That distance places it far enough from the main part of the scablands to make it seem detached, hard to fit into the story of Glacial Lake Missoula and its floods. A big glacial moraine partially blocks the north end of the coulee, about 2 miles north of U.S. 2. The ice that left the moraine there totally plugged the north end of the coulee. All that makes it tempting to relegate Moses Coulee to some earlier ice age, if for no other reason than to evade its puzzles. The problem would be easier to solve if we knew the exact age of Moses Coulee. We do not. But the talus slopes along its walls and the degree of weathering of the basalt provide some evidence of age, vague and crude to be sure, but still helpful.

Talus slopes fringe the bases of cliffs in the scablands, indeed of cliffs everywhere winters are cold. Some people call talus sliderock, some call it scree. By whatever name, it forms when water expands as it freezes within cracks in rock outcrops and pries pieces loose. Those broken chunks of rock tumble to the base of the outcrop, where they accumulate as a rubbly and unstable talus slope. Those slopes grow as the years pass, so their size provides a rough guide to their age. That is an inexact assessment to be sure, more of an intuitive feeling, but still it is helpful. And since talus slopes are loose rubble that torrential floods would easily sweep away, we can be sure that those in the scablands coulees grew after the last great floods passed.

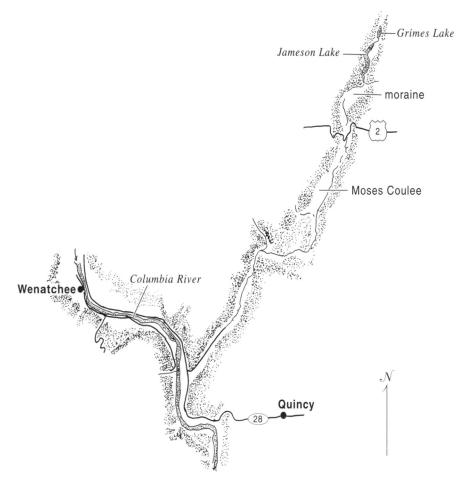

Moses Coulee —Adapted from Ritzville Quadrangle, 1:240,000, U.S. Geological Survey, 1953

Had you walked along the coulees of eastern Washington soon after the Glacial Lake Missoula floods passed, you would have seen raw basalt cliffs that dropped sheer to the floor of the coulee, the floods having cleared all talus. The talus slopes that now fringe the bases of those cliffs grew since then. Those in Moses Coulee are roughly comparable in size to those in Grand Coulee and most other scabland coulees. So far as we can judge from the size of their talus slopes, all those coulees appear to be about the same age.

The degree of weathering of rocks also gives an inexact but never-theless useful impression of how long they have been exposed. The rocks exposed in Moses Coulee are no more weathered than those in Grand Coulee, or most other coulees. Indeed, they are hardly weathered at all, still quite fresh.

So, neither the size of their talus slopes nor their degree of weathering suggest that Moses Coulee is older than Grand Coulee and the other scabland floodways. The evidence does not permit us to relegate Moses Coulee to some earlier ice age. It belongs to the most recent ice age and is definitely part of this story. Perplexing as it seems at first thought, the moraine that blocks the northern part of Moses Coulee makes a good starting point for a simple scenario that links Moses Coulee to Glacial Lake Columbia and Grand Coulee.

Basalt columns stand above a talus slope in Moses Coulee.

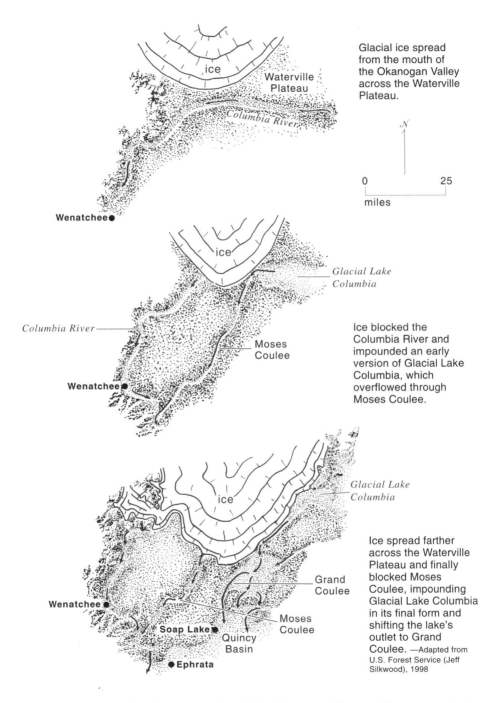

Glacial ice spread from the mouth of the Okanogan Valley across the Waterville Plateau.

ice

Waterville Plateau

Columbia River

N

0   25
miles

Wenatchee●

ice

Glacial Lake Columbia

Columbia River

Moses Coulee

Wenatchee●

Ice blocked the Columbia River and impounded an early version of Glacial Lake Columbia, which overflowed through Moses Coulee.

ice

Glacial Lake Columbia

Grand Coulee

Wenatchee●

Soap Lake●

Quincy Basin

Moses Coulee

●Ephrata

Ice spread farther across the Waterville Plateau and finally blocked Moses Coulee, impounding Glacial Lake Columbia in its final form and shifting the lake's outlet to Grand Coulee. —Adapted from U.S. Forest Service (Jeff Silkwood), 1998

How ice spreading from the south end of the Okanogan Valley could have impounded an early version of Glacial Lake Columbia that drained through Moses Coulee, then a later version that drained through Grand Coulee

As the most recent ice age approached its maximum, ice from British Columbia moved south down the deep Okanogan Valley and into Washington. It spread from the mouth of the Okanogan Valley across the flat plain of the Waterville Plateau like thick pancake batter spreading slowly across a griddle. The spreading ice blocked the Columbia River to impound an early version of Glacial Lake Columbia, which overflowed through Moses Coulee, along with torrents of Glacial Lake Missoula floodwater. All that water entered the Columbia River a few miles south of Wenatchee. The floods scoured the valley walls and deposited giant ripples along the Columbia River below the point where the flow entered from Moses Coulee.

As the ice spread farther across the Waterville Plateau, it filled the northern part of Moses Coulee and deposited the big moraine about 2 miles north of U.S. 2. The ice and its moraine blocked Moses Coulee, displacing the outlet of Glacial Lake Columbia east to its final position, a stream valley destined to become Grand Coulee. After that happened, Moses Coulee continued to carry glacial meltwater, but no more catastrophic floods. Ice deeply filled Moses Coulee above the moraine, where Jameson and Grimes Lakes are. The little road along the floor of the coulee winds for several miles past the moraine. Watch for its lumpy hills of glacial till generously littered with boulders.

The moraine in the northern part of Moses Coulee is one small part of the long ridge of glacial till, the Withrow moraine, that records the farthest reach of the ice of the most recent ice age. Another part of the same moraine perches on the hills just west of Grand Coulee. Had the most recent ice age lasted longer and its ice spread farther, it would have blocked Grand Coulee, just as it had earlier blocked Moses Coulee. That would have forced a further diminished Glacial Lake Columbia to find a new outlet still farther east, no doubt through one of the spillways to the Telford–Crab Creek scabland.

Most of the flow that maintained Glacial Lake Columbia was probably summer meltwater from the enormous mass of ice that filled the lowlands of British Columbia. The lake also received all the flow from the Coeur d'Alene and St. Joe Rivers, as well as some from several lesser streams. The volume of water that continually drained from Glacial Lake Columbia was fully appropriate to a large river. That flow, with the help of the Glacial Lake Missoula floods, enlarged the original stream valley to the grandiose dimensions of Grand Coulee.

A talus slope and basalt cliffs rise above Jameson Lake in Moses Coulee north of U.S. 2.

The overflow from Glacial Lake Columbia eroded through the dark basalt lava flows at the north end of Upper Grand Coulee and bit into the pale granite beneath. Fractures in granite are generally more widely spaced than those in basalt, so they define much larger blocks than those in basalt. And its fractures do not break granite into neat vertical columns that easily separate. The granite at the head of Upper Grand Coulee maintained the overflow from Glacial Lake Columbia at a stable level that was too low to float its ice dam. Had that spillway been on basalt, it would have eroded much deeper, Glacial Lake Columbia would have been much shallower, and the scablands would not be as they are.

Pale granite beneath dark basalt near the north end of Upper Grand Coulee

Waves along the north shore of Glacial Lake Columbia lapped in some places onto glaciers that filled deep mountain valleys, in other places against the rocky ridges between those valleys. It was a splendid scene. Bedrock in the ridges includes large areas of granite and various kinds of metamorphic rocks, as well as smaller areas of Belt rocks. The much lower south shore followed the low and undramatic drainage divide south of the present Spokane River. Rocks there are flood basalt flows still lying nearly as flat as when they poured red-hot and glowing across the Columbia Basin.

The lower ends of the glaciers along the north shore of Glacial Lake Columbia surely floated in the lake and broke off in pieces that became icebergs. Had you visited the low south shore of the lake, you would have seen great flotillas of icebergs drifting before the wind and in the current flowing west to Grand Coulee. A large proportion of the rocks frozen in those icebergs were boulders of granite and metamorphic rock. When the icebergs finally ran aground and melted somewhere downstream, they left those boulders as distant souvenirs of Glacial Lake Columbia.

The effect of Glacial Lake Missoula dumping into Glacial Lake Columbia was like that of an elephant taking a flying leap into a motel swimming pool. Glacial Lake Columbia sloshed back and forth with the impact, massively overflowing with every heroic slosh. Anyone paddling a kayak on the lake would have been washed right over its low southern edge and into the scablands.

How deep was Glacial Lake Columbia? An overflow channel survives above Coeur d'Alene Lake, where Glacial Lake Columbia spilled across the divide at the head of Lake Creek, in Idaho, and west down the valley of Rock Creek. The base elevation of the channel shows that the great floods filled the Spokane Valley to a depth of at least 500 feet. Add to that the unknown depth of the flow through the overflow channel.

Prominent knobs of black basalt, dozens of them, stand as local landmarks in Spokane. They are isolated remnants of lava flows that were continuous until the great Glacial Lake Missoula floods scoured the floor of the Spokane Valley. Think of them as monuments to the extraordinary power of the currents that flowed through the deeply submerged Spokane Valley every time Glacial Lake Missoula drained.

Watch along Interstate 90 between Post Falls and Spokane for broad benches of coarse flood gravel exposed in many roadcuts. They are the downstream end of the flood gravel deposits that cover most of the floor of Rathdrum Prairie. Glacial Lake Missoula floods flushed most of the gravel through Rathdrum Prairie and onto the floor of the Spokane Valley. Glacial meltwater pouring out of Rathdrum Prairie added more as the glaciers of the most recent ice age rapidly melted.

In 1982, J. G. Rigby was a student working on his master's thesis in the Hangman Valley a few miles south of Spokane. The location is important because that valley was a quiet backwater of Glacial Lake Columbia where sediment could accumulate in peace, sheltered from strong currents. Rigby described a series of at least twenty layers that range in thickness from 3 feet to as many as 17 feet. Each has gravel at its base and grades upward into fine sediment in the manner typical of flood deposits. The apparent flood deposits lie sandwiched between sequences of glacial lake sediment complete with varves. The lake sediment accumulated during the regular routine of Glacial Lake Columbia. The flood deposits arrived suddenly when

Knob of basalt in Spokane —Sandra Alt photo

Glacial Lake Missoula emptied into the lake. The geologic picture resembles that at the mouth of Ninemile Creek west of Missoula, where deposits of river silt are sandwiched between sequences of glacial lake sediments (chapter 6).

In 1984, Brian Atwater of the U.S. Geological Survey reported the results of his painstaking study of the sediments deposited in Glacial Lake Columbia. He found thick sections of glacial lake sediments, long sequences of dark and light varves, that faithfully record the seasons of the lake's existence. Within those sequences, he found what appear to be flood deposits from Glacial Lake Missoula. Atwater concluded that the record he saw was consistent with the record of floods in the Palouse River valley, preserved in the Touchet formation (Chapter 25). Both of these records are also consistent with the record of the fillings and emptyings of Glacial Lake Missoula exposed in roadcuts at the mouth of Ninemile Creek. Those deposits and many others all seem to tell the same story, but some geologists disagree.

V. R. Baker and R. C. Bunker argued in 1987 that some, perhaps most, of the floods in the scablands came from Glacial Lake Columbia, instead of Glacial Lake Missoula. J. Shaw and his numerous colleagues further pursued that point of view in 1999. Those geologists contend that the water came from British Columbia into Glacial Lake Columbia, which then overflowed catastrophically across the scablands and down the Columbia River. Testing that idea against the evidence will undoubtedly be one of the major concerns of future investigations in the scablands.

18

THE GRANDADDY SCABLANDS
The Cheney-Palouse and
Telford–Crab Creek Floodways

THE DRAINAGE DIVIDE that now defines the southern edge of the valleys of the Columbia and Spokane Rivers west of Spokane was near the southern shore of Glacial Lake Columbia. Water spilled across its low points when Glacial Lake Missoula floods suddenly overfilled Glacial Lake Columbia. The main overflow was from the Spokane Valley, where a large proportion of the floodwater entering from Rathdrum Prairie poured southwest and eroded the Cheney-Palouse scabland. Farther west, the floods spilled south over the central part of the divide to erode the Telford–Crab Creek scabland. The rest of the floodwater, still an enormous volume, emptied through Grand Coulee at the western end of Glacial Lake Columbia.

The water pouring through the Cheney-Palouse and Telford–Crab Creek scablands ran roughshod over the Palouse Hills, which are huge heaps of windblown silt, probably dunes. During some dry time in the past, perhaps during several dry times, wind from the southwest swept the silt across an older landscape eroded in the basalt lava flows. The silt buried that older landscape so completely that hardly a visible trace survives. The thickness of the windblown silt varies widely, from where it does not quite bury the tops of the old hills to where it fills the old valleys as much as several hundred feet deep. Pine trees grow on the basalt of the old hilltops.

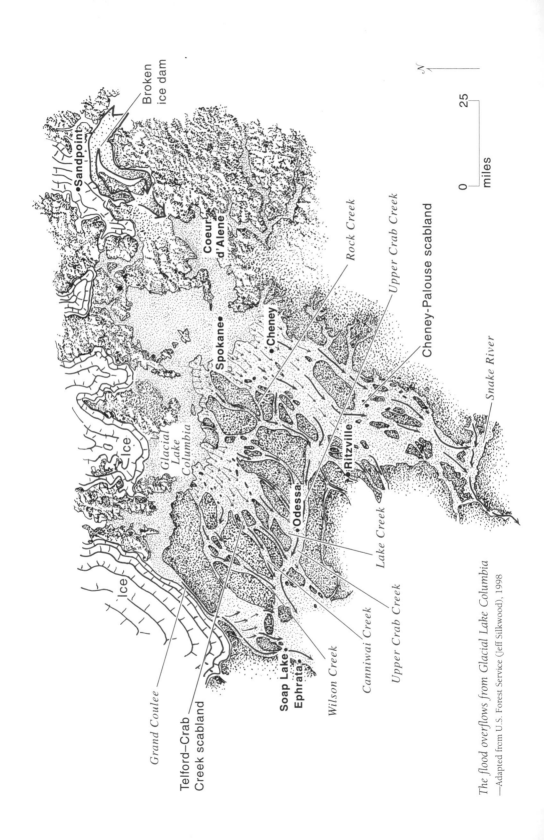

Broken ice dam

Sandpoint

Coeur d'Alene

Grand Coulee

Telford–Crab Creek scabland

Ice

Ice

Glacial Lake Columbia

Spokane

Cheney

Rock Creek

Upper Crab Creek

Cheney–Palouse scabland

Odessa

Ritzville

Lake Creek

Snake River

Soap Lake
Ephrata

Wilson Creek

Canniwai Creek

Upper Crab Creek

N

0   25

miles

The flood overflows from Glacial Lake Columbia
—Adapted from U.S. Forest Service (Jeff Silkwood), 1998

Watch the occasional roadcuts in the gray Palouse silt for the intricate patterns of thin layers of silt that record the shifting winds of thousands of years ago. The wind still blows that loose silt in whirlwinds that dance high across the Palouse Hills on hot summer afternoons. Sometimes the dust fills the air with choking clouds that convert refreshing summer showers into great splatterings of mud.

Lewis and Clark and their party nearly starved for lack of wild game as they trekked through the darkly forested mountains of central Idaho. They were overjoyed to emerge from the forest and into the open expanse of the Palouse Hills, which supported an abundance of wild game. Now the Palouse Hills nourish consistently marvelous crops of grain, lentils, and peas, all highly nutritious. The Palouse silt makes wonderfully fertile soil.

It is not clear just when the Palouse silt blew in, but it was certainly in place in time for the great Glacial Lake Missoula floods. The floods eroded hundreds of those silt dunes out of the Cheney-Palouse and Telford–Crab Creek scablands and spread that fertile silt across the floors of all the temporary lakes that filled and drained with the floods.

The Cheney-Palouse scabland is the classic, the grandaddy of all scablands. It is a broad expanse of channels that branch and join over and over again to weave a complexly braided pattern across an expanse some 90 miles long and 15 to 25 miles across. The massive floods dropped about 1,300 feet in elevation in their headlong rush toward the Palouse River. The weave of channels is extremely dense and closely intertwined at its northeastern end, with small areas of high ground between the channels. The pattern opens to the southwest as both the individual channels and the areas of high ground between them become larger.

Interstate 90 passes through the northern edge of the Cheney-Palouse scabland along most of the route between Spokane and Sprague Lake. Watch for numerous dry channels, many with ponds and small lakes in their floors. Knobs of basalt bedrock, scrubbed bare of soil, rise between the channels. Pine trees flourish, as they do wherever they can sink their roots into the basalt.

Kolks plucked the bedrock, one columnar block of basalt after another, to erode basins of all sizes and in numbers beyond counting. The basins are now dry depressions, ponds, and several respectable lakes. Sprague Lake, the largest visible from Interstate 90, fills a basin in the floor of a large floodway. According to P. Weis and W. Newman of the U.S. Geological Survey, the flow that eroded Sprague Lake's basin was 8 miles wide and 200 feet deep. A number of other large lakes exist well beyond the view of Interstate 90. Some of the secondary roads south of Interstate 90 wind their tortuous way through deep valleys large enough to hold substantial rivers but now dry except for occasional lakes and ponds in their floors. Deeply weathered columnar basalt in cliffs near the rest area on the westbound lane suggests that great floods may have flowed that way during earlier ice ages.

Interstate 90 passes several surviving islands of Palouse silt within the maze of scabland channels. They look like typical Palouse Hills, complete with farms, except that they lie within a surround of scablands. Those island remnants of the Palouse Hills leave no doubt that Palouse silt did indeed cover the entire area of the Cheney-Palouse scabland before the great floods passed through.

The torrents of water that eroded the Cheney-Palouse scabland found their way into the Pasco Basin through several routes. Some poured west through the web of coulees that cut through the Palouse Hills. Some started down the valley of the Palouse River, spilled over its south valley wall to the Snake River, then headed west into the Walla Walla Valley and on into the Pasco Basin. We will consider those routes in chapters to come.

On the spirited occasions of the Glacial Lake Missoula floods, Glacial Lake Columbia also spilled south across the divides at the heads of Wilson, Lake, and Canniwai Creeks, down their valleys, then west down the valley of Upper Crab Creek. Those overflows eroded the Telford–Crab Creek scabland. It broadly resembles the Cheney-Palouse scabland except that its overflows were more widely distributed along a much longer length of shoreline. The Telford–Crab Creek scabland is much less famous that the Cheney-Palouse scabland because it cuts through a thinly populated part of eastern Washington where roads carry mainly local traffic. Washington 21 crosses some of the channels between Wilbur and Odessa.

The channels in the Telford–Crab Creek scabland contain chains of lakes and ponds in bedrock basins that the floods plucked from the basalt bedrock. Small dams now impound many of the higher and larger channels, creating irrigation reservoirs for the Columbia Basin Irrigation Project.

Floodwater coming out of the Telford–Crab Creek scabland flowed straight west down the wide boulevard of Upper Crab Creek and into the Quincy Basin. Upper Crab Creek also carried floodwater west from the Cheney-Palouse scabland through the channels of Coulee Creek, Deep Creek, and Rock Creek. The floods reamed the valley of Upper Crab Creek to a size much larger than anything the present modest stream could possibly erode.

The flow emptying west from Upper Crab Creek dumped a giant gravel bar across the northern Quincy Basin, almost to Soap Lake. Long trains of giant ripples on the gravel leave no doubt that it was dumped under very fast water, presumably the great floods. Washington 28, between Odessa and Soap Lake, provides excellent views of the oversize Crab Creek Valley and its deposits of flood gravel.

J Harlen Bretz pointed out in 1923 that many of the smaller Palouse islands in the Cheney-Palouse and Telford–Crab Creek scablands have a distinctive teardrop shape, with the sharp end of the drop facing into the flow like the bow of a ship. And most wear a faint but distinct high water mark, or high flood mark, that looks like a flange closely fitted around their sides.

Imagery returned from spacecraft orbiting Mars reveals an enormous canyon that cuts almost half the way across the planet, a canyon far larger than any on Earth. V. R. Baker and his colleague D. J. Milton pointed out in 1974 that the eroded canyon floor on Mars looks very much like the big scablands in eastern Washington. In particular, the intertwined patterns of the scabland channels and the teardrop shapes of the Palouse islands look very much like the channels and rock islands in the floors of the Martian canyons. It was a fertile suggestion. That broad similarity now leads many planetary scientists to conclude that floods at least as catastrophic as those that eroded the scablands once swept across Mars.

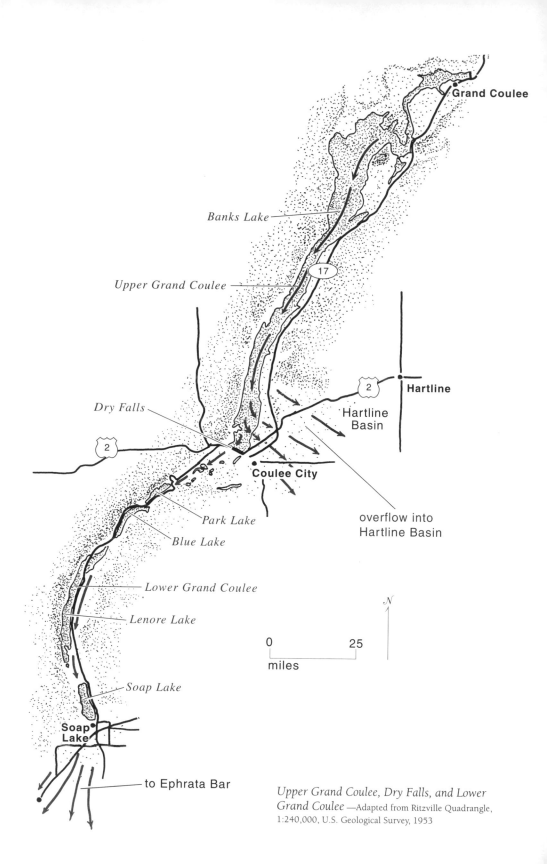

Grand Coulee

Banks Lake

Upper Grand Coulee

17

2 **Hartline**

Dry Falls

Hartline
Basin

2

Coulee City

Park Lake

overflow into
Hartline Basin

Blue Lake

Lower Grand Coulee

Lenore Lake

N

0   25
miles

Soap Lake

**Soap
Lake**

to Ephrata Bar

*Upper Grand Coulee, Dry Falls, and Lower
Grand Coulee* —Adapted from Ritzville Quadrangle,
1:240,000, U.S. Geological Survey, 1953

REGULATING GLACIAL LAKE COLUMBIA
Grand Coulee

THE WESTERNMOST AND perhaps the greatest spillage from Glacial Lake Columbia ran south through Grand Coulee. It poured directly into Upper Grand Coulee, across Dry Falls, and then down Lower Grand Coulee into the Quincy Basin. The resistant granite at the head of Upper Grand Coulee prevented the overflow from eroding a spillway deep enough to drain Glacial Lake Columbia.

Grand Coulee was probably a perfectly ordinary stream valley until it began taking the routine overflow from Glacial Lake Columbia and occasional torrents of floodwater. Together, these flows enlarged the original valley into grand dimensions. The relative contribution of regular overflow from Glacial Lake Columbia and Glacial Lake Missoula's floods is not entirely clear, but Grand Coulee does look like a scabland channel. The floods probably played the major role.

Upper Grand Coulee begins near Grand Coulee Dam, about 550 feet above the Columbia River. That elevation tells us the depth of Glacial Lake Columbia at its point of overflow. It also explains why the Columbia River did not flow through Grand Coulee after Glacial Lake Columbia drained. Upper Grand Coulee continues south about 25 miles to Dry Falls, varying in width between 1 and 6 miles.

A spectacular wall of basalt that rises between 800 and 900 feet above the floor of Upper Grand Coulee bounds its west side along most of its length. The wall follows the east side of a sharp fold called the Coulee monocline. Here, the basalt lava flows tilt sharply down to the southeast at an average angle of about 45 degrees. That flexing shattered the brittle basalt, making it much more vulnerable to erosion than it would otherwise have been. The Hartline Basin east of Upper Grand Coulee is the broad trough that accompanies the sharp arch.

The much lower east wall of Upper Grand Coulee is only about 200 feet high at its southern end near Coulee City. Glacial Lake Missoula floods swept right across it and dumped tremendous volumes of sediment in a broad gravel bar that spreads across the flat country of the Hartline Basin. The water drained south to the Quincy Basin.

Grand Coulee Dam impounds Franklin D. Roosevelt Reservoir on the Columbia River just above Upper Grand Coulee. The dam was built primarily to store irrigation water for the Columbia Basin Irrigation Project, which brings water to the fertile deposits of Palouse silt farther south. A powerhouse generates electricity, some of which drives enormous pumps that lift water from the reservoir into Banks Lake, the high starting point for the project.

Banks Lake fills Upper Grand Coulee between the low North Dam and the equally unassuming Banks Dam at the lake's southern end. It feeds water into an elaborate system of canals, ditches, and secondary reservoirs that carry it to all parts of the irrigation project. The lake is shallow, confined as it is between low dams, so it does not fill the coulee, but just covers its floor. Some talus slopes at the base of the cliffs are now submerged, but the rest remains almost as the great floods left it.

The pale granite exposed around Grand Coulee Dam and in the northern end of Upper Grand Coulee makes a far more reassuring foundation for a large dam than would the more closely fractured basalt.

Water spilling south from Glacial Lake Columbia flowed down Upper Grand Coulee, then thundered over Dry Falls to drop more than 350 feet into Lower Grand Coulee. Dry Falls is a complex of dry cataracts in a row of scalloped cliffs more than 3 miles wide. That is more than five times the width of Niagara Falls. The rumble of the falls would have been audible from miles away, and the tremendous power of the water surely made the ground tremble. Great clouds of spray rose from the falls in all seasons, and in winter made an icy wonderland as it froze onto every rock and tree.

The high west wall of Upper Grand Coulee rises above Banks Lake.

Dry Falls. The spread of plunge pools over a large area below the falls shows that the floods did not merely plunge vertically over the precipice. The water went over the edge in a broad hump that carried much of it some distance from the precipice. —D. W. Hyndman photo

Imagine how easy it would be to stand at the edge of a cliff of columnar basalt, kick pieces off the columns, and watch them tumble onto the talus slope below. The rush of water just as easily ripped blocks of columnar basalt from the cliff at the lip of Dry Falls, then swept them downstream. So the falls retreated upstream as the floods swept blocks of basalt from its lip, leaving Lower Grand Coulee in its wake. No one knows exactly how fast that happened, but the 15 or so miles of Lower Grand Coulee were eroded during the years between the time when ice plugged Moses Coulee and when Glacial Lake Columbia finally drained. The upstream retreat of Dry Falls more nearly resembled Bonaparte's famous retreat from Russia than the kind of orderly withdrawal now so tediously in progress at such trifling features as Niagara Falls.

Water dropping over a falls is moving very fast when it reaches the streambed below, and it lands hard. The falling water erodes a deep

View downstream from Dry Falls, down the canyon the falls left behind as they retreated —D. W. Hyndman photo

plunge pool where it lands, and the water generally fills the old plunge pool with gravel as the falls retreat upstream. The several ponds beneath and just downstream from Dry Falls are its last plunge pools.

Artists' renderings of Dry Falls in action generally look like magnified versions of Niagara Falls, with a white sheet of water dropping vertically over the edge of the cliff. That may have been the view when Grand Coulee was carrying only the routine overflow from Glacial Lake Columbia. But the picture changed when the Glacial Lake Missoula floods overflowed through Glacial Lake Columbia and down Grand Coulee. Then the flow across Dry Falls was so deep that it probably looked more like a sharp slope in the torrent than an ordinary waterfall. And the water was so muddy with silt eroded from the scablands upstream that it looked more like cocoa than the standard white of postcard waterfalls. We will see that mud again as flood deposits in the broad valleys farther downstream.

The precipice that once thundered under a large river and occasional great floods is now still. It looks south across the remains of its old plunge pools, now a maze of quiet lakes and ponds in the floor of the canyon beneath. If Banks Dam were to fail, the lake's water would go over Dry Falls in a deeply disappointing repeat of past events.

Lower Grand Coulee is the canyon that Dry Falls left as it retreated upstream. As always with streams, the strongest flow followed the outside of the bend, so the floods hugged the broad arc of the Coulee monocline down Lower Grand Coulee. That concentrated the erosive power of their flow against the fold, which was also where the rocks were most fractured and least resistant to erosion. The deep channel the floods eroded along the broad arc of the Coulee monocline now holds a series of bedrock lake basins crowded against the west wall of Lower Grand Coulee.

It is tempting at first thought to suppose that those lakes are old plunge pools left behind as the lip of Dry Falls retreated upstream. But if they are old plunge pools, why did the continuing flow not fill them with gravel? It seems more likely that something in the fluid dynamics of the flow caused persistent erosion in those spots. Perhaps they mark places where the flow generated large numbers of kolks as it sheared past the steep wall of the coulee.

Whatever their origin, the lakes in Lower Grand Coulee make marvelous scenery. Washington 17 passes through one dramatic landscape after another as it follows the chain of Park, Blue, Alkali,

The west wall of Lower Grand Coulee. The basalt ledges tilt in the Coulee monocline.

View in Lower Grand Coulee

Lenore, and Soap Lakes. Beautiful as they are, their water is alkaline, the common fate of lakes without outlets in dry regions.

Floodwater exiting Lower Grand Coulee dumped gravel across an expanse of country from south of Soap Lake, past Ephrata, all the way to the Frenchman Hills. The northern part of that enormous deposit is coarse gravel, which feathers southward into a broad fringe of sand and silt that stayed in suspension long after the gravel settled out. Washington 17 crosses broad benches of coarse gravel littered with boulders between Ephrata and Moses Lake.

Many people call this gravel deposit the Ephrata Fan, which is unfortunate because it is not an alluvial fan. Trains of giant ripples here and there on its surface leave no doubt that it was laid down beneath a flow of very fast water. The elevations of the dry channels across Babcock Ridge on the western side of the Quincy Basin show that the gravel was deposited under as much as 300 feet of water. The deposit is actually a giant gravel bar—better to call it the Ephrata Bar.

If any great quantity of water had drained from Lower Grand Coulee after the floods deposited the Ephrata Bar, it would have cut a large channel through the bar. None exists. Evidently, Glacial Lake Columbia emptied and the Columbia River simultaneously took the drainage at about the time of the last great flood. Grand Coulee carried no more water.

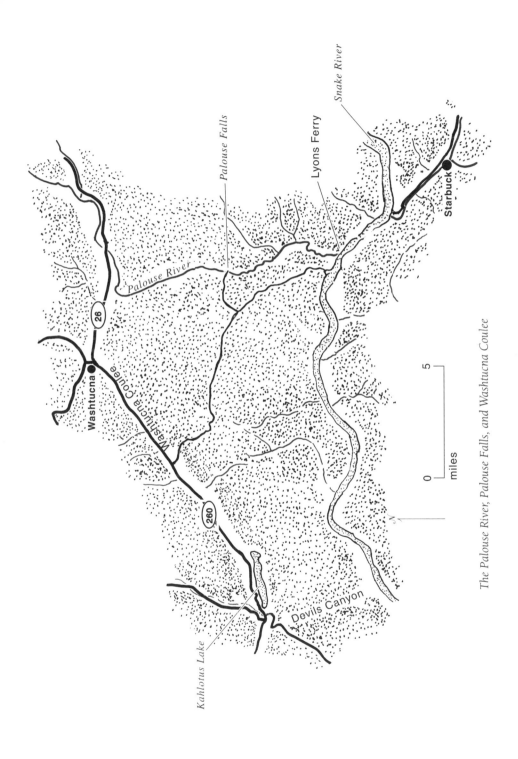

The Palouse River, Palouse Falls, and Washtucna Coulee

THE SHORTCUT TO
THE SNAKE RIVER
The Palouse River, Palouse Falls,
and Washtucna Coulee

THE CHENEY-PALOUSE SCABLAND ends where most of the floodwater roared into the old valley of the Palouse River, hellbent for the Snake River and the Pasco Basin. A large portion of that torrent spilled over the south valley wall and down a flood channel to Palouse Falls. Then it continued down the canyon below the falls to the Snake River and finally to the Pasco Basin. That was a shortcut to the Snake River. The rest of the flood continued west down the lower valley of the Palouse River, now Washtucna Coulee. It flowed through that coulee to join Esquatzel Coulee at Connell, and flowed south into the Pasco Basin.

The lower Palouse River still leaves its original valley to follow the flood channel shortcut over Palouse Falls and on to the Snake River. The early floods probably dropped directly into the Snake River. Their successors eroded the lip of that waterfall, working it upstream to its present position. As the waterfall retreated, it left in its wake the deep canyon of the Palouse River. Monumental Dam on the Snake River now floods the lowest part of that canyon.

The road into Palouse Falls State Park passes old channels considerably above the channel that carries the modern Palouse River. And they look much more weathered, much older, than the big

Palouse Falls

flood channel that carries the modern river. Many geologists suspect these and similar channels elsewhere are relics of Glacial Lake Missoula floods during one or more ice ages before the most recent ice age. No one knows when those floods may have happened, if indeed they did happen.

Palouse Falls is truly a bizarre spectacle that makes a strong case for Bretz. The modern Palouse River is hardly more than a creek, much too small to have eroded the large valley that carries it to Palouse Falls. Nor could the modern river have eroded the deep canyon below the falls. Those features are clearly products of a far greater flow. And Palouse Falls themselves drop into a ridiculously oversized plunge pool much too large to fit the modern river. This, too, is a relic of the heroic flows of the past. Evidently, the flow of the puny modern stream is too weak to erode the lip of the falls and leave that old plunge pool behind.

J. C. Gilluly was one of the more severe of the critics who attacked Bretz in 1923. He continued his barrage of criticism as he rose to become a prominent member of the U.S. Geological Survey and one

View down the canyon below Palouse Falls

of the authors of a widely used introductory geology textbook. Many years after the great conflict with Bretz began, he finally ventured into the scablands and visited Palouse Falls. The story has Gilluly taking a long look, then asking how anyone could have been as wrong as he had been. Of course, it is easily possible to be so wrong if you make up your mind without having considered or even seen the evidence. Despite his epiphany at Palouse Falls, Gilluly's textbook continued to insist that geologists should not appeal to catastrophic causes in their attempts to explain events of the geologic past.

Washtucna Coulee is the abandoned valley of the Palouse River below its point of diversion to Palouse Falls and one of the more spectacular sights in the scablands. The floods reamed it out to a size that could carry a river far larger than the modern Palouse River. The old tributaries of the Palouse River still empty into Washtucna Coulee, but they do not contribute enough water to maintain a flow. Scoured basins near its lower end hold a few ponds of alkaline water.

Washtucna Coulee is remarkably straight because its high south wall follows the steep north flank of the Saddle Mountains ridge. The

Scabby outcrops of basalt in Washtucna Coulee

much lower north valley wall was submerged during the larger floods. Both the valley walls and the floor of Washtucna Coulee are very rough with scabby outcrops of dark basalt poking through the dry brush.

The modern Snake River empties into the Columbia River at Pasco, but it did not during Glacial Lake Missoula floods. The enormous surges of water that arrived on those drastic occasions reversed the flow of the Snake River and backed it upstream more than 100 miles. The reverse flow ran past Lewiston and far up the valley of the Clearwater River in Idaho. Those backwards floods left large gravel bars high on the valley walls of the Clearwater River, complete with pebbles typical of those in the Snake River.

Layers of pale silt of the Touchet formation sandwiched between darker Snake River gravels exposed in a roadcut on Washington 261 near the Snake River, just west of Starbuck

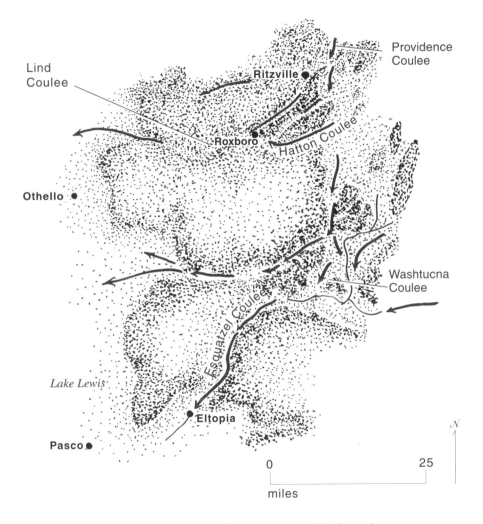

Lind
Coulee

Providence
Coulee

Ritzville ●

Roxboro ●

Hatton Coulee

Othello ●

Washtucna
Coulee

Esquatzel Coulee

Lake Lewis

Eltopia ●

Pasco ●

N

0   25

miles

*Coulee routes from the Cheney-Palouse scabland into the
Pasco Basin* —Adapted from U.S. Forest Service (Jeff Silkwood), 1998

THE GATHERING OF THE WATERS
Beyond the Cheney-Palouse Scabland

ANOTHER LARGE FLOW of Glacial Lake Missoula's floodwater followed Washtucna Coulee and several lesser routes into both the Quincy and Pasco Basins. Their map is a confusing web of stream valleys and coulees, vivid testimony to a volume of water that totally overwhelmed the carrying capacity of the existing drainage. Water filled the lowlands, followed old stream channels where they were handy, and cut new channels as it needed them.

Part of the water that flowed west from the Cheney-Palouse scabland entered Lind Coulee east of Ritzville and followed it through a broad tract of Palouse Hills to the Quincy Basin. Washington 21 follows Lind Coulee between Lind and Warden to Lower Crab Creek.

The floods went through Lind Coulee in a foaming torrent of brown water that suddenly filled it, then poured through for at least a week, perhaps several. Although the floods certainly left a wasteland in their wake, Lind Coulee is now a broad and fertile valley with very few scabby outcrops of basalt in its walls. Farm fields spread across most of its flat floor, which is filled with flood deposits. Canals carry irrigation water in parts of Lind Coulee, but the natural stream that flowed through it before the floods is gone.

The part of Lind Coulee near Roxboro is famous in limited geologic circles for its giant ripples. They are among the most spectacular in eastern Washington, between 10 and 12 feet high and about 250 feet from crest to crest. Faintly visible but distinct scour marks on the valley walls about 350 feet above the floor of the coulee record

129

Giant ripples make the odd pattern on the valley floor of Lind Coulee.

the depth of the flow. Like many tracts of giant ripples, these are hard to see at close hand. Roads up the sides of the coulee provide higher vantage points and better views.

Most of the floodwater that rushed through Lind Coulee continued straight west down Lower Crab Creek, through the broad valley between the Frenchman Hills and Saddle Mountain ridges to the Columbia River. That water continued down the river into the Pasco Basin. The rest of the floodwater joined the southward flow over the eastern end of the Saddle Mountains ridge and into the Pasco Basin. That water helped erode the Othello Channels across the Saddle Mountains ridge.

U.S. 395 follows Providence and Hatton Coulees south from the area south of Ritzville to Connell. That is where Esquatzel Coulee gathered the water flowing south through Providence and Hatton Coulees with that flowing west out of Washtucna Coulee. The highway follows the east side of Esquatzel Coulee south from Connell to Eltopia.

Watch west from U.S. 395 between Connell and Eltopia for good views of Esquatzel Coulee. Along most of the way, it is a broad and sharply defined channel obviously capable of carrying a magnificent torrent. Imagine it filled to the brim with a rush of turbulent water with enough Palouse silt suspended in it to give it the color of strong cocoa. Esquatzel Coulee fades near Eltopia, where its water finally flowed into Lake Lewis.

The south valley wall of Lower Crab Creek

THE GRAND EMPTYING
Floodwater Exits the Quincy Basin

THE MAIN FLOOD route out of the Quincy Basin was straight west through the broad valley of Lower Crab Creek to the Columbia River. Some of that water had spilled south over the submerged eastern end of the Frenchman Hills ridge, eroding the Drumheller Channels on its way. The rest was emptying from the western end of Lind Coulee. Water that did not follow the Crab Creek route poured south over the submerged eastern end of the Saddle Mountains ridge, eroding the Othello Channels. Meanwhile, water spilling over Babcock Ridge at the western edge of the Quincy Basin eroded the scabland channels at Crater Coulee, Potholes Coulee, and Frenchman Spring Coulee.

Lower Crab Creek flows almost straight west, following a course along the base of the steep north face of the Saddle Mountains ridge. That is the lowest part of the valley. Its floor rises gently to the north, up the long Royal Slope to the Frenchman Hills ridge on the horizon about 9 miles north.

Lower Crab Creek enters the Columbia River just above the spectacular gorge through the Saddle Mountains, a few miles south of Wanapum Reservoir. The Columbia River was in its present course before the folded arch of the Saddle Mountains began to rise across its path some millions of years before the first Glacial Lake Missoula flood. The river's flow was vigorous enough to erode the gorge through the ridge as it rose.

The nearly straight and very steep north face of the Saddle Mountain ridge rises south of Lower Crab Creek and the road down Lower

133

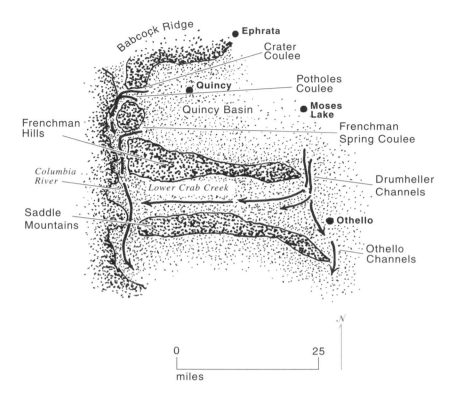

The Lower Crab Creek route to the Columbia River —Adapted
from Walla Walla Quadrangle, 1:250,000, U.S. Geological Survey, 1963

Crab Creek. Its long row of nearly vertical cliffs rises 1,000 feet and
more and trends almost directly west. The lava flows of dark basalt
exposed in the cliffs make ledges that slant down to the south—the
opposite direction to what we would expect on the north flank of a
folded arch. In this case, the lava flows on the north side of the fold
tilted beyond vertical and rolled over until they were slightly over-
turned.

The passage of the great floods cleared all traces of alluvial fans
and talus slopes from the base of the steep cliffs along Lower Crab
Creek. Those we now see grew during the last 13,000 or so years of
weathering and erosion. They seem small, considering the height and
steepness of the cliffs. That is also true of most other scabland channels, but somehow it seems more striking along Lower Crab Creek.

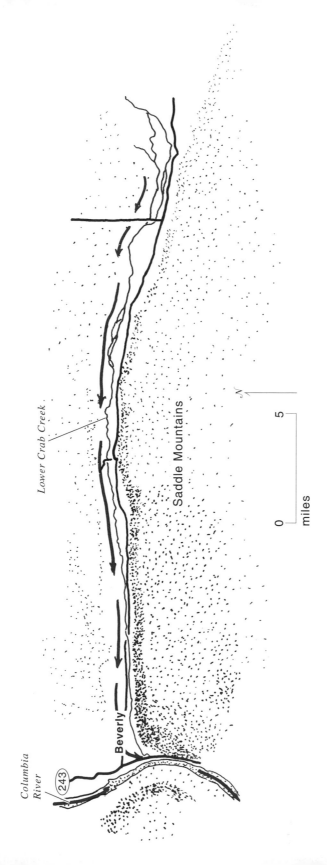

Columbia
River
243

Beverly

Saddle Mountains

Lower Crab Creek

0   5

miles

Water routes out of the Quincy Basin

Patches of dunes that sparkle in the sun at the west end of Lower Crab Creek are white because they consist largely of quartz grains. That presents a problem because the basalt lava flows of eastern Washington do not contain quartz. The Columbia River brought the quartz sand from the country north and east of the great lava field, where the rocks contain plenty of quartz. Then the wind blew it off the river floodplain and swept it up into dunes. The dunes are much younger than the great Glacial Lake Missoula floods and not part of their story.

The main flood route south from the Quincy Basin to the Pasco Basin passed between the east ends of the Frenchman Hills and Saddle Mountains ridges and a large tract of Palouse Hills farther east. That was a narrow strait for so much water, especially considering that the extreme eastern ends of the two ridges were submerged. They made broad sills on the bottom that greatly obstructed the flow over them. The rush of water across the submerged east ends of the two ridges eroded the Drumheller and Othello Channels.

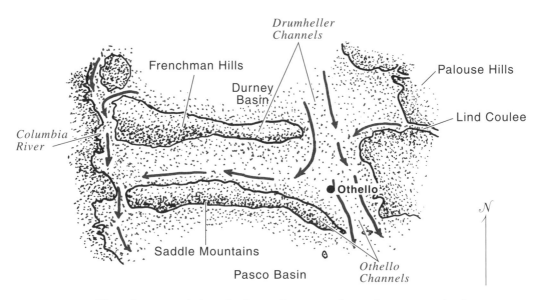

Water flowing south from the Quincy Basin passed over the eastern ends of Frenchman Hills and the Saddle Mountains ridges and eroded the Drumheller and Othello Channels. —Adapted from U.S. Forest Service (Jeff Silkwood), 1998

The Drumheller Channels cover approximately 50 square miles of the low eastern end of the Frenchman Hills. J Harlen Bretz, who knew something about scabland channels, walked through them with his students. He considered them the most complex and bewildering part of the scablands.

The floods first eroded through a deposit of soft sediment called the Ringgold formation, then cut as much as 300 feet into the hard basalt beneath. The water level in the Quincy Basin must have been substantially higher than that in the Pasco Basin to provide a gradient steep enough to drive such a fiercely erosive flow. So the Drumheller Channels provide further evidence that the Quincy Basin filled with water much faster than the Pasco Basin—further evidence, if any were needed, that Glacial Lake Missoula emptied suddenly.

The easiest place to see a small part of the Drumheller Channels is from the road along the top of O'Sullivan Dam, which impounds Potholes Reservoir. The view west from the east end of the dam shows the smooth rise of the uneroded Frenchman Hills above the level of the floods, where the Ringgold formation is still in place. The view south reveals the horribly ragged landscape in the lower part of the Drumheller Channels.

Potholes Reservoir floods the Drumheller Channels north of O'Sullivan Dam, so they are no longer visible. Canals carry excess irrigation water from the north into Potholes Reservoir and some individually dammed scabland channels. Other canals redistribute the water south into the Pasco Basin.

The Othello Channels, about 8 miles south of Othello, are a lesser version of the Drumheller Channels. Lesser because much of the water that eroded the Drumheller Channels flowed west down Lower Crab Creek, instead of continuing south over the eastern end of the Saddle Mountains ridge. Small dams impound the higher elevations of the Othello Channels to store irrigation water for distribution to the Pasco Basin.

The highest exits from the Quincy Basin were across the crest of Babcock Ridge, which defines the western border of the basin. It reaches from the Frenchman Hills ridge in the south to the Beezley Hills in the north. The crest of Babcock Ridge stands about 300 feet above the basin floor.

Floodwater spilled west out of the Quincy Basin across Babcock Ridge through Frenchman Spring Coulee in the south, Potholes Coulee in the middle, and Crater Coulee in the north. The overflows for all three are at nearly the same elevation. The water must have suddenly burst over the crest of the ridge, then poured through cascades and over waterfalls down the long slope west to the Columbia River.

The Frenchman Spring Coulee is just north of Interstate 90, about 4 miles southwest of George. It includes a classic dry waterfall that overlooks the Columbia River. Potholes Coulee is about 6 miles south of Quincy and about 3 miles west of Washington 281. The flow through its channel continued down the western slope of Babcock Ridge and over two waterfalls, each about 200 feet high, into the Columbia River. The two waterfalls dropped over two basalt ledges because the upper one eroded more rapidly than the lower one, and so the upper falls retreated faster and farther. A broad bench developed between them as they retreated 2 miles upstream from their original position at the river. Washington 28 passes Crater Coulee just west of Quincy. This coulee carried water west from the Quincy Basin to spill over a spectacular waterfall into the Columbia River.

Irrigation water impounded in a coulee on the crest of Babcock Ridge

Small dams now impound all three of the coulees on the crest of Babcock Ridge, as well as some of the channels on its western slope. They store water for the Columbia Basin Irrigation Project. But the view up the slope from the Columbia River remains almost unchanged. The dry waterfalls below Crater Coulee are especially impressive from that perspective.

The little roads on the crest of Babcock Ridge offer occasional glimpses west down the channels that furrow the long slope to the Columbia River and east across the great emptiness of the Quincy Basin. Water does not flow uphill, so the only way it could enter any of those channels is from the Quincy Basin, and that could happen only if the basin was filled to a depth of at least 300 feet. The floods came very quickly indeed if they filled the Quincy Basin to overflowing before the water could drain through the broad passage down Lower Crab Creek—more evidence that Glacial Lake Missoula emptied catastrophically.

Moses Lake and the sand dune field that impounds it are not really a legitimate part of this story. Both formed long after the last Glacial Lake Missoula floods passed, probably at least several thousand years after, and are only marginally related to them. But there they are, right in the middle of the scablands, and much too big to ignore.

The fine outer fringe of sand and silt around the southern part of the Ephrata Bar extends to the long ridge of the Frenchman Hills. Those fine sediments provide splendid fodder for the strong westerly winds that blow clouds of silt into the Palouse Hills and sweep the lagging sand into tracts of dunes. Most of the dunes have irregular shapes, but some are sharply defined crescents with the tips of their horns leading the way northeast, in the direction they are moving.

Some geologists, most notably V. R. Baker, suggest that Lower Crab Creek adopted a channel that the last drainage from Lower Grand Coulee eroded across the Ephrata Bar. Old topographic maps drawn long before the O'Sullivan Dam impounded Potholes Reservoir clearly show that the wind drove the dunes across the creek, which did not carry enough flow to maintain a clear channel. So Lower Crab Creek and its tributary Rocky Ford Creek backed up behind the dunes to fill the two largest arms of Moses Lake. It is one of the few natural

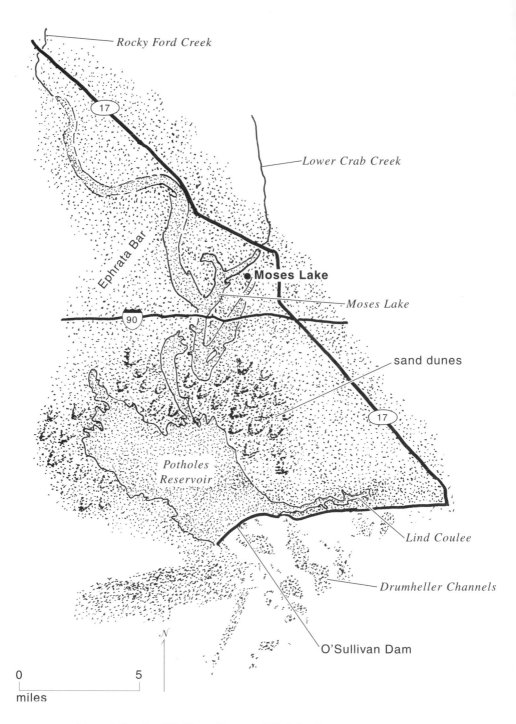

Rocky Ford Creek

Lower Crab Creek

17

Ephrata Bar

● **Moses Lake**

Moses Lake

90

sand dunes

17

Potholes
Reservoir

Lind Coulee

Drumheller Channels

O'Sullivan Dam

N

0   5

miles

Moses Lake, the O'Sullivan Dam, and Potholes Reservoir —Adapted from
Ritzville Quadrangle and Walla Walla Quadrangle, 1:250:000, U.S. Geological Survey, 1953

lakes in eastern Washington that does not flood a scabland basin plucked from the hard basalt bedrock.

As Moses Lake filled, it raised the water table in the dune field, creating ponds and marshes in the low areas between dunes. They have been slowly filling ever since as the wind blows sand into them, but not out. Potholes Reservoir submerged part of the dune field, converting the tops of many of the higher dunes into the little sandy islands that dot the southwestern part of the reservoir.

Large areas of the Moses Lake dune field now sleep under a stabilizing blanket of grass and brush. Obviously, the climate is wetter now than it was when the entire dune field was marching before the wind. Abundant evidence from many places shows that climates drier than ours have prevailed at times since the end of the most recent ice age, most famously about 8,000 years ago. That is probably when blowing sand blocked Crab Creek and impounded Moses Lake.

Active sand dunes south of Moses Lake. The sand owes its medium gray color to the dark minerals from basalt.

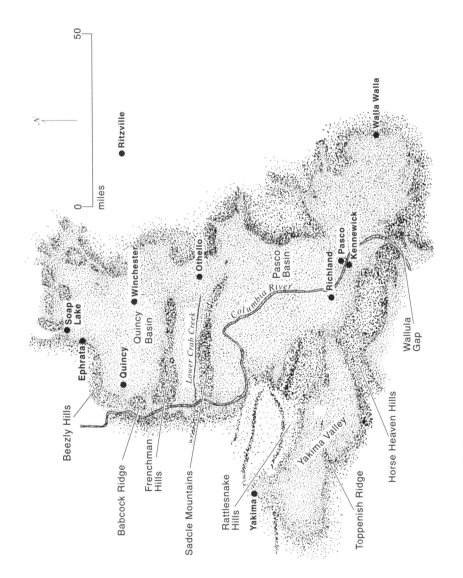

The basins of Lake Lewis —Adapted from U.S. Forest Service (Jeff Silkwood), 1998

Beezly Hills

Babcock Ridge

Frenchman Hills

Saddle Mountains

Rattlesnake Hills

Yakima

Toppenish Ridge

Yakima Valley

Horse Heaven Hills

Wallula Gap

Ephrata

Quincy

Quincy Basin

Soap Lake

Winchester

Othello

Lower Crab Creek

Columbia River

Pasco Basin

Richland

Pasco

Kennewick

Ritzville

Walla Walla

N

miles

0

50

23

A PATCHWORK OF VALLEYS
Lake Lewis

EVERY DROP OF WATER that sloshed out of Glacial Lake Columbia and poured across eastern Washington eventually found its way into an enormous temporary lake north of the Horse Heaven Hills. Geologists call it Lake Lewis after Meriwether Lewis, who with William Clark led their Corps of Discovery down the Columbia River in the fall of 1805 and back the next spring.

The main part of Lake Lewis flooded the Quincy and Pasco Basins, which are north and south, respectively, of the lesser ridges of the Saddle Mountains and Frenchman Hills. At its fullest, the lake reached west into the Yakima Valley, east into the Walla Walla Valley, and north in the Quincy Basin to somewhere beyond Ephrata. It extended about 120 miles from east to west and about 60 miles from south to north. Lake Lewis flooded more than 2,000 square miles and was almost 800 feet deep at the site of the Tri-Cities.

W. T. Symons first described Lake Lewis in 1882 in his geologic report about the Columbia Plateau. He and other early geologists assumed it had been a permanent lake that spilled through a low saddle in the Horse Heaven Hills and drained as its overflow eroded Wallula Gap. Many years later, geologists finally realized that the Columbia River began eroding Wallula Gap millions of years before Lake Lewis first filled with water. This is another place where the river eroded its bed fast enough to keep pace with a steep ridge rising across its path. Lake Lewis was never more than a temporary body of

143

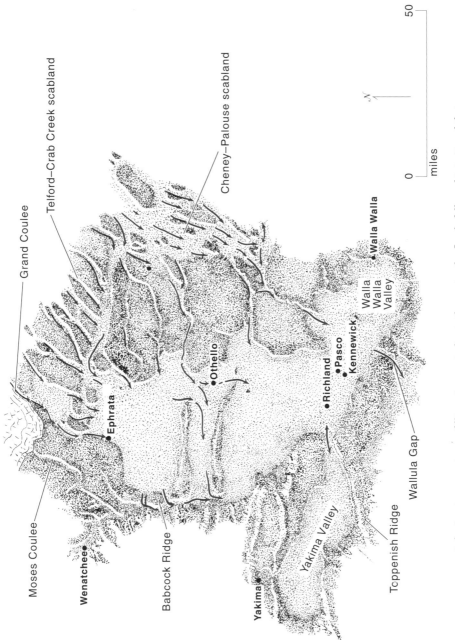

Grand Coulee

Telford–Crab Creek scabland

Cheney–Palouse scabland

Moses Coulee

Wenatchee●

Babcock Ridge

Yakima●

Yakima Valley

Toppenish Ridge

Wallula Gap

●Ephrata

●Othello

●Richland

●Pasco
●Kennewick

Walla
Walla
Valley

●Walla Walla

N

0   50

miles

Lake Lewis at maximum filling. Arrows indicate the routes the floods followed into it and their route out through the Wallula Gap —Adapted from U.S. Forest Service (Jeff Silkwood), 1998

water that filled with each Glacial Lake Missoula flood, then drained through Wallula Gap.

The Walla Walla Valley was the easternmost part of Lake Lewis. The Walla Walla River empties into the Columbia River north of the Horse Heaven Hills, just below its junction with the Snake River at Pasco. The higher fillings of Lake Lewis flooded an area of the Walla Walla River valley about 24 miles from east to west and 18 miles from north to south, as much as 300 square miles. At least part of the site of Walla Walla was then submerged under as much as 50 feet of water.

The Yakima Valley is at the western extremity of the former Lake Lewis. It lies north of Toppenish Ridge, which is the western

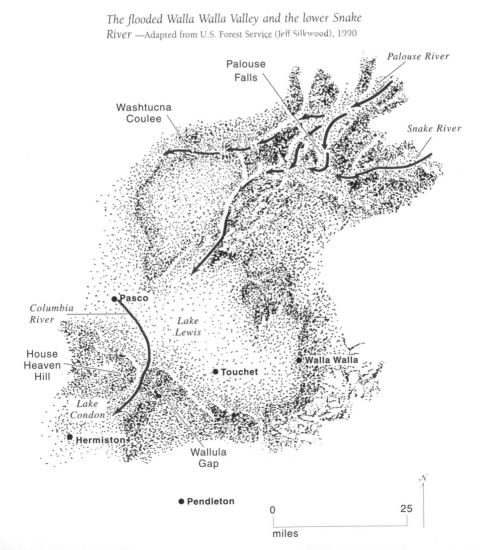

The flooded Walla Walla Valley and the lower Snake River —Adapted from U.S. Forest Service (Jeff Silkwood), 1990

Palouse River

Palouse Falls

Washtucna Coulee

Snake River

Columbia River

●Pasco

Lake Lewis

House Heaven Hill

●Walla Walla

●Touchet

Lake Condon

●Hermiston

Wallula Gap

N

●Pendleton

0   25

miles

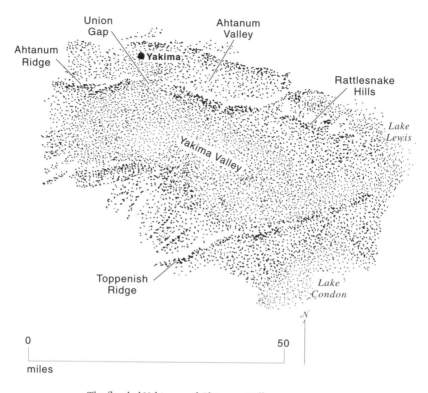

The flooded Yakima and Ahtanum Valleys —Adapted from
U.S. Forest Service (Jeff Silkwood), 1998

continuation of the Horse Heaven Hills, and south of the Rattlesnake
Hills and Ahtanum Ridge. The town of Yakima is in the much smaller
Ahtanum Valley, which is north of the Yakima Valley, and connects
with it through Union Gap. The Yakima River eroded Union Gap as
the geologically continuous ridge of the Rattlesnake Mountains and
Ahtanum Ridge rose across its path. When Lake Lewis reached its
higher levels, it flooded approximately 600 square miles of the Yakima
Valley and a much smaller area in the slightly higher Ahtanum Valley.
Part of the site of Yakima was under water. All that water entered and
left through the open east end of the Yakima Valley.

The only exit for all the water temporarily ponded in the various
parts of Lake Lewis was through Wallula Gap, just south of Pasco.
Lake Lewis probably lasted only a few weeks after each filling, perhaps only a few days after its lesser fillings.

24

THE ARCHIVES OF THE FLOODS
The Touchet Formation

RICHARD FOSTER FLINT of Yale University was the most prominent of the geologists who denounced Bretz's theory of a catastrophic flood in the scablands. And he became Bretz's most energetic and persistent detractor. Flint also wrote the most widely used glacial geology textbook of its time. In his later years, he came to consider himself the great guru of the ice ages. A good many geologists privately referred to him as "The Pope of the Pleistocene." His discussion of the scablands dwindled in successive editions of his textbook to a single sentence as his efforts to discredit Bretz failed. That created an unfortunate gap in a generally excellent book, a great loss for all the students who used it.

Still, it must be said to his great credit that Flint closely studied the scablands in the field as he tried to devise explanations that did not involve sudden catastrophic floods. In the course of that work, he described and named the Touchet formation from outcrops near Touchet in the valley of the Walla Walla River, about 14 miles west of Walla Walla. That is one of the places where its soft layers are nicely exposed.

The Touchet formation is a stack of pale sedimentary layers that range from a few inches to several feet thick. They are mostly silt, with some sand and gravel. Close examination shows that the silt came mainly from the Palouse country to the north, doubtless eroded from the Cheney-Palouse and Telford–Crab Creek scablands. Some of the pebbles are bits of rock that could have come only from Glacial

147

Lake Missoula, others from Glacial Lake Columbia. The individual layers—and the formation itself—are generally thicker in the areas where Lake Lewis was deepest.

Each layer in the Touchet formation is coarsest at the base and becomes progressively finer upward. Geologists call these graded beds; they are typical of flood deposits. To make your own graded bed, put some mixed sediment in a jar, add water, shake it, then put the jar on a table and let the sediment settle. The heaviest particles will settle first and the lightest last, so the deposit in the bottom of the jar will grade from coarsest at the base to finest at the top. That is essentially what happens when sediment settles out of a muddy flood to the floor of a lake.

R. J. Carson and his colleagues described sequences of layers in the Touchet formation with all the typical characteristics of flood deposits in 1978. But they did not conclude that each of those layers records an individual Glacial Lake Missoula flood. R. B. Waitt of the

Touchet beds south of Touchet, Washington —D. W. Hyndman photo

*Richard Foster Flint
in 1937* —Courtesy
Lincoln Washburn

U.S. Geological Survey studied the outstanding exposures of the Touchet formation in Burlingame Canyon, in the valley of the Walla Walla River about 8 miles west of Walla Walla. Burlingame Canyon is an enormous gully that escaping irrigation water eroded many years ago. Waitt concluded in 1980 that the graded layers in the Touchet formation do indeed record a series of floods, at least forty-one of them. That is five more than the number of fillings of Glacial Lake Missoula that R. L. Chambers and I found near the mouth of Ninemile Creek west of Missoula. It is easy to imagine that the last five fillings of the lake did not rise to an elevation high enough to leave a record at the site in western Montana.

The thickest layers of graded sediment exposed in Burlingame Canyon are at the base of the stack, and they become progressively thinner upwards. Some geologists interpret that pattern as evidence that each successive flood found less Palouse silt to erode in the Cheney-Palouse and Telford–Crab Creek scablands. But the pattern of flood deposition that Waitt found in Burlingame Canyon corresponds to the pattern of lake deposition in the Glacial Lake Missoula sediments at Ninemile Creek. Both suggest that each successive flood was smaller than the one before, and therefore less erosive as it passed through the scablands.

Geologists first recognized and described the Touchet formation in the floor of the temporary Lake Lewis. Closely similar deposits exist wherever the great floods ponded. Only the names of these deposits change from place to place. All make extraordinarily productive soil, eminently worth irrigating. The Palouse silt is as fertile where it was redeposited as at its original site.

I believe that both lake and flood deposits show with great clarity that Glacial Lake Missoula filled and emptied at least several dozen times during the most recent ice age. If so, the scablands are the product of dozens of great floods, not of the single flood that J Harlen Bretz proposed in 1923, nor of the seven or nine floods he envisioned some 30 years later. If the glaciers of earlier ice ages impounded earlier versions of Glacial Lake Missoula, the probable total number of floods becomes much greater than the several dozen of the most recent ice age.

It seems genuinely ironic that R. F. Flint, who was the most fervent of Bretz's many opponents, described and named the Touchet formation, in which many geologists would later find evidence of dozens of catastrophic floods.

THE WAY WEST
Through Wallula Gap and into Lake Condon

U.S. 730 FOLLOWS a spectacular route through Wallula Gap, beneath its scoured walls of basalt in somber shades of dark gray. They rise above the quiet waters of Lake Wallula, impounded behind McNary Dam on the Columbia River. The floods that roared through this dark canyon as Lake Lewis drained inundated vast reaches of Oregon and a smaller area of Washington as they backed up behind the bottleneck of the Columbia Gorge. Geologists call that enormous pool of stalled floodwater Lake Condon to honor Thomas Condon, one of the great pioneers of Oregon geology.

John E. Allen devoted a large part of a long career to the Glacial Lake Missoula floods, especially to their passage down the Columbia River. He and Marjorie Burns published a book, *Cataclysms on the Columbia*, that summarizes much of the story of the Glacial Lake Missoula floods for the lay reader. They estimated that Lake Condon covered approximately 1,500 square miles, probably for a period of a week or more during each flood—long enough to let much of the mud settle out of the murky water.

Watch for the great sheets of flood gravel visible from Interstate 84 near Boardman, Oregon, just below Wallula Gap. The same strong currents that swept that gravel through Wallula Gap and into Lake Condon also scoured the basalt lava flows. Their scattered remnants, mostly north and east of Hermiston, closely resemble the basalt monuments in Spokane and likewise testify to the tremendous erosive power

151

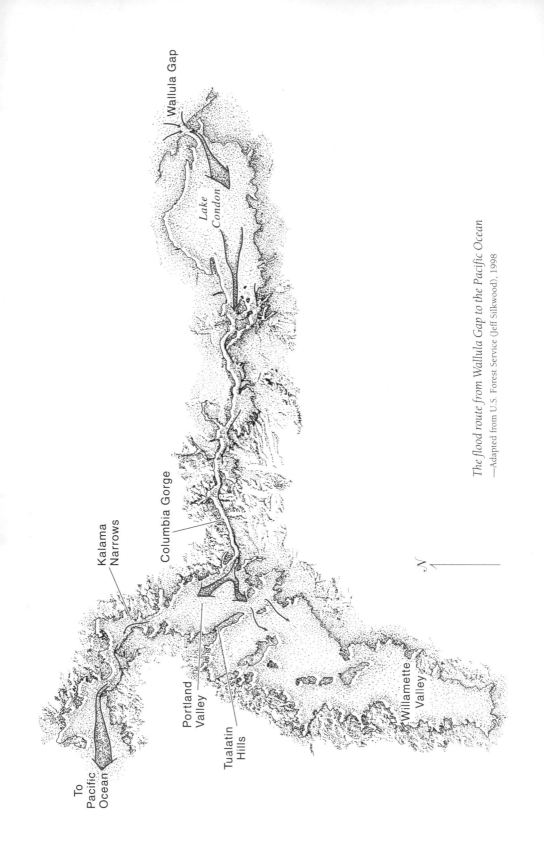

Wallula Gap

Lake Condon

Columbia Gorge

Kalama Narrows

Portland Valley

Tualatin Hills

Willamette Valley

To Pacific Ocean

N

The flood route from Wallula Gap to the Pacific Ocean
—Adapted from U.S. Forest Service (Jeff Silkwood), 1998

of flood currents tearing across columnar basalt. The most famous of these black monoliths is Hat Rock, the centerpiece of Hat Rock State Park. It stands on the south bank of the Columbia River a few miles northeast of Hermiston, and it really does look like the battered high crown of an enormous old stovepipe hat.

In its days of drainage glory, Wallula Gap was a narrow passage between Lake Lewis and Lake Condon, a constriction within a muddy inland sea hellbent for the Pacific Ocean. But Wallula Gap did not

Wallula Gap —Adapted from the Walla Walla and Pendleton Quadrangles, 1:250,000, U.S. Geological Survey, 1953

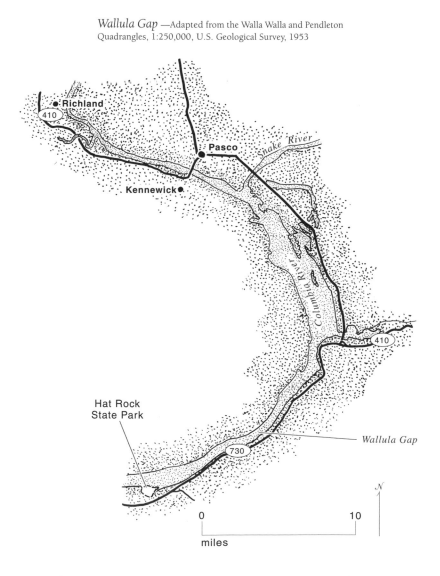

Wallula Gap in a view downstream along U.S. 730

govern the drainage of Lake Lewis. The Columbia Gorge was the effective bottleneck that controlled floodwater levels in both lakes. Water poured through Wallula Gap to feed Lake Condon as its water drained through the Columbia Gorge.

An indistinct line between scrubbed bedrock cliffs below and slopes still heavily mantled with soil above records the upper level of the strong flows through Wallula Gap. Using that line as a starting point, V. R. Baker concluded that the peak discharge through Walulla Gap was about 1.66 cubic miles per hour. That discharge would have occurred only during the largest floods and could not have lasted long even then. At all times, the flow through Wallula Gap must have been very close to that passing through the controlling bottleneck of the Columbia Gorge.

Like all the temporary lakes along the route of the floods, Lake Condon never stood at one level long enough to leave a shoreline. In the absence of shorelines, the upper level of flood scouring in Wallula Gap provides the best evidence of the maximum surface elevation of Lake Condon, about 1,100 feet. Overflow channels provide further

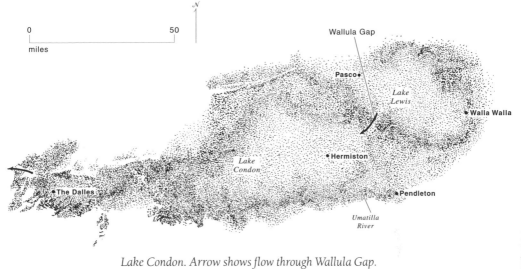

Lake Condon. Arrow shows flow through Wallula Gap.
—Adapted from U.S. Forest Service (Jeff Silkwood0

evidence, but they tell only the base elevation of flow through them, not its depth. Widely scattered erratic boulders that must have arrived in floating icebergs imply minimum lake levels, but the odds are that very few of them, if any, came to rest at the highest lake level. Many of those boulders are granite or gneiss, so they must have come from Glacial Lake Columbia. Many are rounded as we would expect of boulders that had been carried in a glacier.

Flood deposits of silt and sand in graded layers cover the floor of the former Lake Condon. They are like those in the Touchet formation in the floor of Lake Lewis. The soils weathered on those sediments during the last 13,000 years grow wonderful crops where irrigation water is available. The variety and quality of the late summer produce in roadside fruit stands around Hermiston and Irrigon testify to the productivity of relocated, weathered, and irrigated Palouse silt. The melons are famously succulent.

Several overflow channels near the west end of Lake Condon cross the drainage divide south of the Columbia River into the Rock Creek drainage. Rock Creek empties into the John Day River, which empties into the Columbia River. So the water that poured across the divide took a little excursion down Rock Creek and the John Day

River before it rejoined the main body of the floods on their way to the Pacific Ocean.

The most easily accessible of those old spillways crosses the divide at the head of Alkali Creek south of Arlington. Others lie farther west at the heads of Jones, Blalock, and Phillipi Canyons. The base elevations of all four require a lake surface elevation of at least 1,000 feet, but of course that is a minimum figure. If the depth of the overflow was 100 feet, then the lake elevation was about 1,100 feet, the same as the flow through Wallula Gap.

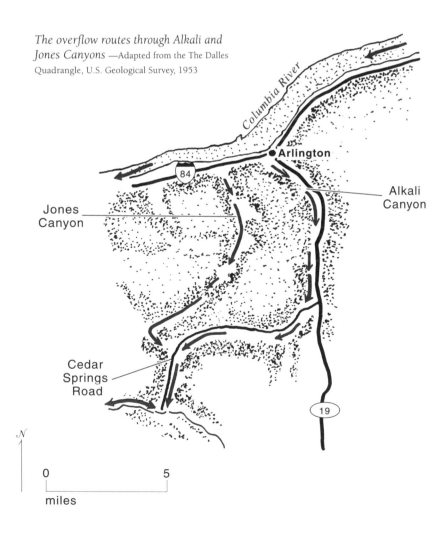

The overflow routes through Alkali and Jones Canyons —Adapted from the The Dalles Quadrangle, U.S. Geological Survey, 1953

Oregon 19 follows Alkali Creek south from Arlington. Cedar Springs Road branches west from it just south of the divide and follows the path of the floods through a coulee so broad and so distinct that it must record the passage of a deep and steady flow over the drainage divide. The coulee descends almost to Rock Creek, which means that water flowing down it reached the creek before water backing up from the John Day River could fill its valley that far upstream. Once again, strong evidence that the water level rose very quickly, that Glacial Lake Missoula really did empty catastrophically.

Ordinarily, the John Day and Deschutes Rivers flow into the Columbia River, but the Glacial Lake Missoula floods temporarily reversed them. If we use the maximum elevation of Lake Condon, about 1,100 feet, as a guide, we can figure that Lake Condon backed the flow of the John Day River as far south as Thirty Mile Creek, about 35 miles south of Interstate 84. And it backed the Deschutes River approximately as far south as Maupin, about 52 miles south of Interstate 84 on U.S. 197.

Knowing that the flood level rose very fast makes it easy to imagine a sudden blast of cold wind blowing up the canyon just ahead of the wall of water that ran aggressively upstream like a fast tidal bore. In a few days the water ran slowly back downstream as Lake Condon drained through the Columbia Gorge.

Interstate 84 between Boardman and the Columbia Gorge offers one spectacular view after another of the valley wall north of the Columbia River. Views become especially outstanding west of The Dalles, where the valley begins to narrow into the Columbia Gorge. Look there for the line between the vigorously scrubbed lower valley wall with its ragged outcrops of flood basalt and the higher slopes still deeply upholstered in soil. The line is fuzzy in most places, fairly clear and sharp in others. It is sharpest in the narrower parts of the valley and along the outsides of curves that deflected the main current against the north wall of the valley. The scrub line is approximately 800 feet above the river, so the floodwaters reached an elevation of approximately 1,100 feet above sea level along the upper part of this reach of the river, closer to 1,000 feet in the lower part.

This view north from Interstate 84 shows the rocky lower valley walls that lost their soil in the floods and the rounded hills above that still have theirs. —John Rimel photo

The scoured basalt makes broad benches and long ledges that display beautiful palisades of vertical basalt columns. The Mary Hill Museum and Stonehenge Replica stand high above the Columbia River on one of those benches, beside U.S. 97, across from Biggs, Oregon. The ledges of dark basalt that continue for miles and miles testify to the enormous size of the basalt lava flows of middle Miocene time.

The floods cleared occasional broad basalt benches near the valley floor, dotted here and there with isolated monuments of basalt, remnants of the lava flow that once lay above them. Fractures that the floods eroded into deep grooves cross many of those benches. Some of the Columbia River flowed through those grooves long after the floods passed. Most are now submerged behind big dams, remembered only through old photographs.

Large slabs of basalt tilt steeply down toward the highway from the cliffs south of Interstate 84 in the stretch between Blalock and Philippi Canyons. They may have dropped after the floods undercut them, but it is impossible to be sure.

ANOTHER TIGHT SQUEEZE
The Columbia Gorge

WHERE THEY SQUEEZED into the narrow bottleneck of the Columbia Gorge, the greatest floods were about 1,000 feet deep, about 500 feet deep as they poured out of the western end of the gorge into the Portland Valley. That steep gradient drove the floodwater through the Columbia Gorge in a ferocious rush. The andesite bedrock of the Columbia Gorge could hardly have been more vulnerable to those torrents.

Most of the volcanic rock in the Cascade Range is andesite. It is easy to recognize: a medium to dark gray rock flecked with very pale gray, white, or faintly greenish crystals of the mineral plagioclase feldspar. Andesite is the typical rock of volcanic chains.

Some andesite erupts as simple lava flows. A larger proportion erupts as ash, some of which lands on the volcano while the rest drifts on the wind to more distant destinations. Large andesite eruptions tend to become rather watery occasions because hot volcanic ash melts snow and ice, and the clouds of steam blowing out of the crater generate thunderstorms. Some of that water mixes with volcanic ash to make mud, which pours down the volcano, picking up rocks along the way. Most andesite volcanoes are big piles of ash and rubbly mudflows knitted together with a few solid lava flows. If there were a building code for mountains, andesite volcanoes would never pass inspection. We could hardly expect the Glacial Lake Missoula floods to leave the Columbia Gorge unscathed.

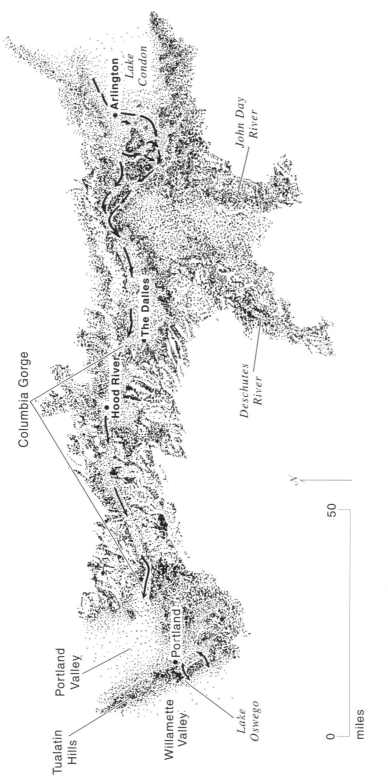

Columbia Gorge

Tualatin
Hills

Portland
Valley

Willamette
Valley

Portland

Lake
Oswego

Hood River

The Dalles

Arlington
Lake
Condon

John Day
River

Deschutes
River

N

0   50

miles

The flood route through the Columbia Gorge —Adapted from U.S. Forest Service (Jeff Silkwood), 1998

An eroded wall of the Columbia Gorge rises almost vertically.
—John Rimel photo

J Harlen Bretz recognized in the early 1920s that the floods not only scrubbed the soil off the lower valley walls of the Columbia Gorge but also deeply eroded its volcanic bedrock. That may seem like an easy observation in the informed view of hindsight, but it could not have been obvious at the time. The Columbia Gorge is too heavily forested to permit many easy views of its rocks, and it is much too big to permit easy perception of its form.

The floods gouged the lower valley walls of the Columbia Gorge deeply on both sides of the river, but with very different effects north

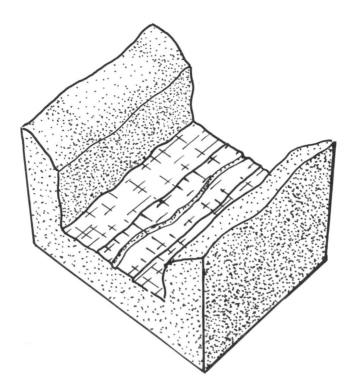

The Columbia Gorge as it was immediately after the last Glacial Lake Missoula floods passed through. Before the floods came, the valley walls sloped gently down to the river. After their passage, the drastically eroded valley walls ended in very steep slopes.

and south. On the south side, streams that enter the gorge tumble down its steepened lower slope in countless waterfalls and cascades, especially when the snow is melting in the spring and early summer. Multnomah Falls, a favorite scenic turnout on Interstate 84, is the largest and best known of these waterfalls. The water drops about 400 feet through a vertical slot it eroded since the last floods passed.

On the north side of the Columbia Gorge, volcanic rocks lie on the Eagle Creek formation. It is mostly volcanic ash, full of petrified wood in many places, and tilted gently down to the south. This soft and sloping formation makes a precarious foundation for the heavy volcanic andesite lying on it, a slippery banana peel under the rocks above. Ever since the eroding floods steepened the lower part of the north valley wall, volcanic rock has been sliding south on the Eagle

View west into the Columbia Gorge with its reamed lower valley —John Rimel photo

Multnomah Falls
—John Rimel photo

Creek formation in a series of giant landslides. Their movement flattens the steepened slopes as it carries their lower parts toward the river. That explains why no waterfalls tumble down the north side of the Columbia Gorge.

Just as ordinary current ripples a few inches across are easier to perceive than giant current ripples a few hundred feet across, so small landslides are easier to recognize than giant ones. That is why those landslides along the north side of the Columbia Gorge are hard to see from either side of the river. But they do exist through approximately 50 square miles of the Columbia Gorge between Washougal and Bonneville.

One of those slides dammed the Columbia River at Bonneville and impounded a lake about 275 feet deep, which apparently lasted

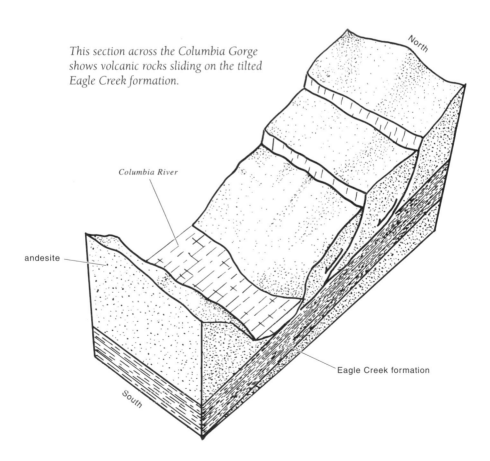

This section across the Columbia Gorge shows volcanic rocks sliding on the tilted Eagle Creek formation.

North

Columbia River

andesite

Eagle Creek formation

South

several years. Then the overflow eroded its spillway across the landslide dam deeply enough to drain the lake. Until that happened, it may have been possible to walk across the river on the slide—the Bridge of the Gods of Indian legend. Radiocarbon dates on wood from trees drowned in the lake show that it all happened about 750 years ago, recently enough to be the plausible basis of a living oral tradition. The Bonneville Dam, which is only about one-third the height of the vanished landslide dam, now floods most of what remained of the Bridge of the Gods. The area of irregular and hummocky topography north of the dam is the lower part of the slide.

Some of the slides along the Washington side of the Columbia River are still moving, still damaging houses and breaking roads, but no one knows if any of them will again dam the river. Landslides move so capriciously that few geologists risk predicting what they will do. A long history of slow movement, or even of no movement, does not guarantee that the tedium will continue into the indefinite future. Most geologists consider a landslide capable of catastrophic movement until it finally reaches the base of its slope.

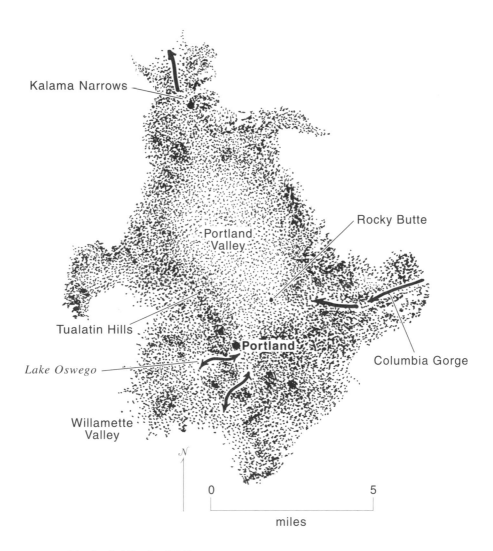

The flooded Portland Valley —Adapted from U.S. Forest Service (Jeff Silkwood), 1998

27

A SUDDEN RUSH OF WATER
Scouring through Portland

As the floods approached the future site of Portland, a steady rumble like distant thunder shook the air for more than an hour. The ground trembled, and its vibration raised a peculiar pattern of little standing waves on the Columbia River. Then a sudden cold wind from the mouth of the Columbia Gorge and a wall of water 500 feet high, dark with muddy sediment, descended upon the region. Perhaps the local people, if any were there, had a few moments to reflect on what their elders had said about floods. But now their apocalypse was upon them.

Urban development has obliterated most relics of the flood around Portland, and the few remaining are very hard to see. Lake Oswego is certainly the most spectacular souvenir of their passage. If present trends continue, it may soon be the only one easily visible.

Rocky Butte, a thoroughly defunct volcanic basalt cinder cone at the north edge of Portland, stood in the path of the floods like a boulder in a streambed. Water rushing against a large rock in a stream commonly erodes the streambed on its upstream side, and that is what happened when the great floods washed against the upstream side of Rocky Butte.

Interstate 205 and Interstate 84 follow the broad trench of the old floodway along the east side of Rocky Butte, although it is now almost impossible to see. Old maps drawn while Portland was young show the eroded floodway and long gravel bars trailing west from both sides of Rocky Butte. The gravel bars enclosed

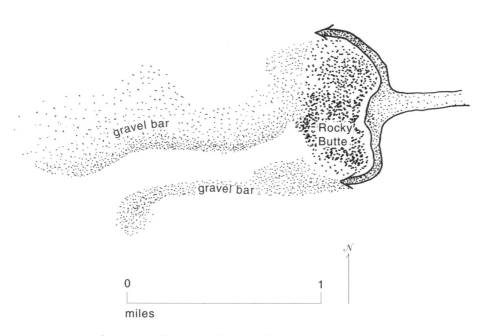

Rocky Butte, with its gravel bars trailing west into the area of the former slackwater downstream —Adapted from Allen and Burns, 1986

a marshy lowland. The geologic picture is very much like that at Locust Hill and Banana Lake, downstream from Rainbow Lake in western Montana.

The valley of the Columbia River is about 1.8 miles wide through its narrows at Kalama, a few miles below Portland. It is astounding that such a broad passage made a fairly effective hydraulic dam when Glacial Lake Missoula's floodwaters arrived. Water fed into the upstream side of the bottleneck faster than it could flow through and out the downstream side. Evidently, the floodwaters arrived very suddenly, even this far below the failed ice dam in northern Idaho. J. E. Allen estimated that about two-thirds of the floodwater continued down the Columbia River, while the other third ponded temporarily in the Portland area and in the Willamette Valley. That third was still an enormous amount of water.

The narrows at Kalama impounded the temporary lake that flooded the Portland and Willamette Valleys, so the water level was about the same in both areas. Allen and Burns estimated the maximum depth of the flow through the Kalama Narrows as about 400 feet, so that was the approximate maximum elevation of the temporary lakes that flooded the Portland and Willamette Valleys.

The long ridge of the Tualatin Hills makes a nearly continuous barrier between the Portland Valley and the Willamette Valley to the south. It is a sharply folded arch that raised flood basalt lava flows to the surface—good plucking rock.

The Willamette River flows through one low pass in the Tualatin Hills ridge, and Lake Oswego lies in the floor of another pass about 5 miles north. The easiest way to explain those passes is to suppose that the Willamette and Tualatin Rivers eroded their channels rapidly enough to maintain their courses while the ridge of the Tualatin Hills rose across them. All the water that flooded the Willamette Valley entered and left through those two narrow gaps.

The flow rushing west through the northern gap that once carried the Tualatin River plucked the basin that holds Lake Oswego. The debris from that basin makes an enormous bar of coarse gravel littered with boulders that spreads about 3 miles. The location of that bar leaves no doubt that water rushing southwest through the gap and into the Willamette Valley eroded the basin of Lake Oswego. And that large gravel bar dumped in its path made it impossible for the small Tualatin River to return to its old course. It now joins the much larger Willamette River south of the Tualatin Hills. If the Tualatin River had been able to resume its old course, it would long ago have filled Lake Oswego with sediment.

A much larger volume of water undoubtedly rushed south through the wider gap that the much larger Willamette River eroded through the ridge of the Tualatin Hills. That flow also dumped a large fan of debris across the flat floor of the Willamette Valley, but that much larger debris dump was too thinly spread to prevent the Willamette River from resuming its old course after the floods had passed. The river promptly filled with sediment any lake basin that was eroded in the gap. That is why we do not see a second and larger version of Lake Oswego near Oregon City, where the Willamette River still flows through its gap in the ridge of the Tualatin Hills.

The flood level in the Portland Valley must have risen very fast indeed to maintain a surface gradient steep enough to drive such a strong flow southwest through the two gaps in the Tualatin Hills. The evidence clearly shows that the Glacial Lake Missoula floods arrived catastrophically, as a wall of water, even this far west. The return flow as the water drained out of the Willamette Valley was not powerful enough to carry large volumes of coarse gravel back into the Portland Valley. No doubt that was because the narrows at Kalama, which were controlling the downstream drainage, let the water level down fairly slowly. The water arrived in a tearing great hurry, then left in a considerably more dignified manner.

The Salem Hills are another ridge nearly parallel to, and geologically similar to, the Tualatin Hills, but farther south. The Yamhill River flows through a broad gap at the north end of the ridge, the Santiam River through another gap farther south. Those two gaps evidently permitted the floods to pass without plucking any basins, indeed without scouring of any kind. That is probably due in part to the width of those two gaps and in part to a slower rise of the water level south of the Tualatin Hills.

THE INLAND BAY
Floods in the Willamette Valley

THE PIONEER OREGON GEOLOGIST Thomas Condon concluded in 1871 that the Willamette Valley had once been a bay, an arm of the ocean that existed at a time of higher sea level. More than a century of further investigation showed that his saltwater bay was actually a series of freshwater lakes that flooded the Willamette Valley as many times as Glacial Lake Missoula suddenly emptied. The floods lasted no more than a few weeks, probably less. They left neither shorelines nor channels eroded across drainage divides to provide us with evidence of the deepest water level, so we must rely on evidence from layers of silty sediment and a scattering of stray rocks.

The main record of the floods in the Willamette Valley is in the layered sediment deposits on the valley floor, the Willamette silt. Ira S. Allison, a distinguished geologist who devoted a long career to the geology of western Oregon, first described the Willamette silt in 1953. J. L. Glenn, who was then a graduate student, provided far more detailed information in 1965, as did R. B. Parsons and his colleagues in 1970. The Willamette silt is almost exactly like the Touchet formation on the floor of the former Lake Lewis in eastern Washington. Some exposures reveal dozens of graded layers stacked to a thickness of more than 100 feet. Like the graded layers in the Touchet formation, each apparently records a separate flood.

Like the Touchet formation, the Willamette silt is yet another deposit of relocated Palouse silt. It spreads the natural fertility of the Palouse Hills across the floor of the Willamette Valley. Think of all

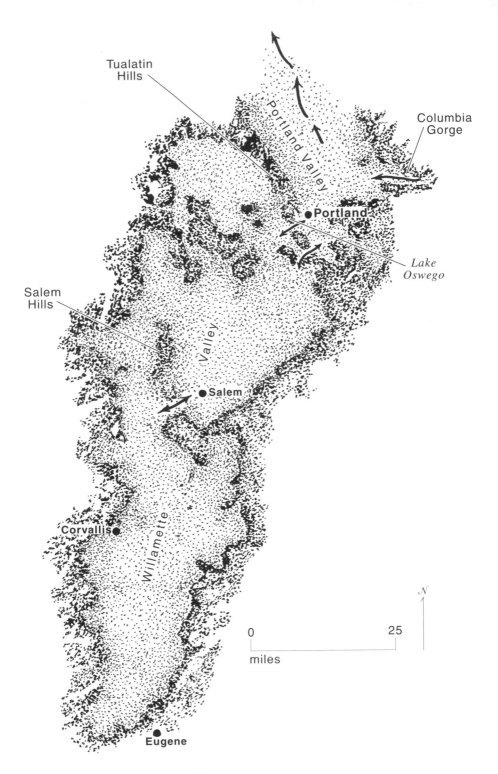

The flooded Willamette Valley —Adapted from U.S. Forest Service (Jeff Silkwood), 1998

the barren bedrock in the Cheney-Palouse and Telford–Crab Creek scablands as you past the lush fields and the flourishing groves of fruit and nut trees in the Willamette Valley.

Erratic rocks provide more evidence for what we know of floods in the Willamette Valley. They are geologically conspicuous because they do not match the material on which they lie, in this case the Willamette silt. They are called erratics because they are out of their natural place. All erratics of whatever size or type were somehow carried from somewhere else to where they now rest. They tell a story. We normally look to erratic boulders for evidence because they are big and easy to see, but erratic pebbles tell the same stories and are much more abundant.

Many of the erratic boulders in the Willamette Valley are angular. If they had rolled under the floods, they would have lost their sharp edges and corners. So something must have picked them up and carried them bodily. Some of the erratic boulders are rounded, but that tells us nothing of how they were moved. They might have been rounded before they started on their long trek to the Willamette Valley. Only angularity counts.

It seems as certain as anything in geology seems that all the erratic rocks of whatever size or degree of rounding in the Willamette Valley arrived frozen inside floating icebergs. Some of the erratic boulders are alone, others lie in groups that commonly contain several quite different kinds of rocks. Either way, they mark places where an iceberg ran aground and melted during the long days of an ice age summer.

Some of the erratic rocks in the Willamette Valley are Belt rocks that could have come only from the northern Rocky Mountains. Most of those rocks were probably embedded in the great ice dam in northern Idaho. Some of the rounded erratic rocks are granite and metamorphic rock that must surely have floated in the icebergs of Glacial Lake Columbia. Whatever their original source, those wayward rocks traveled at least 500 miles.

Ira S. Allison was not one of those who attacked J Harlen Bretz in 1923. He entered the great controversy several years later, and is remembered for having done so with great kindness and tact. He also went beyond mere criticism and developed his own alternative theory, which was probably what motivated him to locate, photograph, describe, and catalog every erratic boulder he could find in the Willamette Valley during the early 1930s. People tend to get rid of stray boulders, so the body of large evidence shrinks every year, gradually reducing to pebbles. Allison's catalog contains all we will ever know of most of the boulders he studied.

Allison's theory invoked a monster ice jam at Portland that made a dam so high and so big that it flooded eastern Washington and the Willamette Valley. In long hindsight, his theory now seems so fanciful and so lacking in supporting evidence that it is hard to imagine why anyone took it seriously, least of all Allison, who was an excellent geologist. But his notion of an ice jam at Portland lingered for decades in a much reduced and somewhat ghostly form. It now seems obvious that the Kalama Narrows below Portland are far more believable as the bottleneck that flooded the Willamette Valley.

The floor of the Willamette Valley rises gently to the south, so the farther south the icebergs drifted with their cargo of erratic boulders, the deeper the flood that brought them there. Reconstructing the

The vanishing Bellevue erratic —D. W. Hyndman photo

flooding of the Willamette Valley became a game of finding the south-ernmost erratic boulders. They seem to be just north of Eugene, nearly at the south end of the valley. A lake high enough to float icebergs that far south was about 400 feet deep. It covered an area of almost 11,000 square miles.

The Bellevue erratic boulder is the largest of them all, an enor-mous chunk of Belt rock that must have come from the northern Rocky Mountains. E. T. Hodge estimated in 1950 that this boulder weighed about 160 tons. Allen and Burns reported in 1986 that it was down to about 90 tons, and shrinking. They made a straight line projection that it would completely disappear by the year 2019— such are the depredations of souvenir hunters. It is now the center-piece of a small state park at Bellevue and within a fence that may delay its final disappearance.

The Willamette meteorite, also a boulder, was found in the woods near West Linn in 1902. Its discovery launched a famous legal dispute over ownership between the man who recognized it as a meteorite and the man on whose land he found it. A long battle that featured many appeals was finally resolved in favor of the land-owner. The meteorite is now in the American Museum of Natural History in New York City. Many years later, geologists realized that the meteorite had been one of a cluster of erratic boulders of several different rock types, exactly the sort of group that commonly marks the last resting place of an iceberg. That makes it seem likely that the meteorite floated into the Portland Hills while frozen in an iceberg.

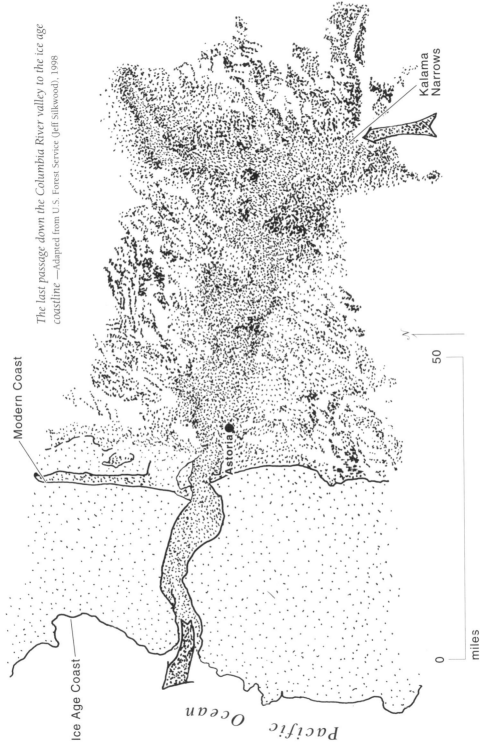

The last passage down the Columbia River valley to the ice age coastline —Adapted from U.S. Forest Service (Jeff Silkwood), 1998

Kalama Narrows

Modern Coast

Astoria

Ice Age Coast

Pacific Ocean

N

0   50

miles

29

THE LAST LAP
On to the Pacific Ocean

THE LAKES THAT TEMPORARILY FLOODED the Willamette Valley soon drained north through the Willamette River, back into the Portland Valley, and on into the Columbia River. No evidence suggests that their draining was nearly as exciting as the sudden rushes of water that filled the Willamette Valley. But they did help maintain a mighty flow of water that scoured the valley of the Columbia River all the way to the ice age coastline. Some of the bedrock along that route is flood basalt, most of the rest is muddy sandstone laid down on the floor of the Pacific Ocean between roughly 60 and 30 million years ago. The basalt is very dark. The oceanic sandstone is dark where freshly broken, but weathers to pale shades of yellowish brown.

After the floods passed through the narrows at Clatskanie, they followed an open valley to the Pacific Ocean, but even here this was an immense flood. The scoured lower valley walls show that the flood crests reached a height of 300 feet at Astoria. Viewpoints along that last stretch of the Columbia River reveal a valley floor much too wide to fit the modern river. The occasional high bluffs are also oversized.

The ice sheets of the most recent ice age stored enough water on the continents to drop sea level more than 300 feet. Fathometer recordings reveal a submerged valley that extends seaward from the mouth of the Columbia River to the ice age coast some 40 miles west of the modern coast. That was where the earliest of the great floods finally reached the ocean. But the climate changed approximately when Glacial Lake Missoula began to fill and empty, and the great

177

glaciers were rapidly melting during most of the years of the lake's existence. Sea level was rising at an average rate of about 100 feet every 1,000 years as the glacial meltwater ran back into the oceans. As sea level rose, the ice age coastline moved rapidly landward. So every great flush entered the ocean a bit closer to the modern coastline. The coast finally reached the site of Astoria hundreds of years after the last Glacial Lake Missoula flood passed.

30

DUSTING OFF THE CRYSTAL BALL
Could It Happen Again?

WE TEND TO FORGET about ice ages amid all the public anxiety over emissions of greenhouse gases and the prospect of the global warming they might cause. What about ice ages?

Our earth has seen a good many ice ages come and go, probably a dozen or more, during the past 2 million or so years of Pleistocene time. All those comings and goings involved widespread, if not global, cooling and warming. Those climatic changes happened through perfectly natural mechanisms that we may or may not understand, except that they surely had nothing to do with human activities. And the years since the most recent ice age ended have brought major changes in climate, periods of much warmer or colder weather and greatly different rain and snowfall than we now know. Some of those changes have happened within the period of written historic record.

The Vikings, for example, attempted to colonize Greenland and North America during a warm climatic period when their North Atlantic sea routes were fairly free of ice. Then the climate cooled, the period of the "little ice age" began, and they were forced to abandon their outposts in Greenland and North America.

It is folly to consider the present climate normal, or to expect it to persist indefinitely. Considered in the climatic context of the last 2 million years, neither global warming nor the onset of a new ice age would be in any way abnormal. Most people would probably regard the coming of a new ice age as environmental degradation on a truly

179

heroic scale, probably much more frightening than an episode of global warming. It could happen.

Although theories abound, no one really knows what causes ice ages. Neither is it entirely clear what ends them, but some evidence suggests that a sudden episode of global warming ended the most recent ice age. Deep ice cores drilled in Greenland contain trapped bubbles of ancient air along their entire lengths. Those bubbles show that a large increase in atmospheric carbon dioxide coincided with the end of the most recent ice age, and seems quite likely to have caused it. Where did the carbon dioxide come from? Possibly from catastrophic breakdown of methane and ice compounds called methane clathrates that exist in enormous volume in sediments on the continental shelves. Once liberated into the atmosphere, methane oxidizes into carbon dioxide.

In any case, we have no reason to assume that the most recent ice age was the last ice age. So far as anyone now knows, a new ice age might start at any time. If the next ice age were to start promptly at noon tomorrow, many thousands of years, perhaps tens of thousands, would pass before enough ice would again fill the valleys of British Columbia to send another great glacier down the Purcell Trench. Then a new ice dam could indeed impound the Clark Fork River at Lake Pend Oreille. New versions of Glacial Lake Missoula could again fill the mountain valleys of western Montana and spill a new series of catastrophic floods across eastern Washington and down the Columbia River.

Yes, it could happen again. But not for a long time, and not without thousands of years of advance warning.

EPILOGUE

SCIENTIFIC CONTROVERSIES eventually end. The great squabble over the scablands should have ended in 1942 when J. T. Pardee showed that Glacial Lake Missoula had indeed emptied catastrophically. It finally came to a ceremonial end in 1962 during a field trip tour of the scablands, part of the Seventh Congress of the International Association for Quaternary Research. Poor health kept J Harlen Bretz at home. At the end of the trip, the participants sent the fading old man in Chicago a telegram reporting their mass conversion to his way of thinking.

It would be pleasant to write that Bretz's detractors saw the light of evidence and reason, and changed their minds. Very few did. The others maintained a hammering drumbeat of criticism for more than another decade. The battle finally wore itself out more through natural attrition of the participants than through general consideration of the evidence. Most of Bretz's critics went to their graves still stoutly maintaining that perfectly ordinary processes of erosion had somehow produced the extraordinary landscapes of the scablands. It is pleasant to report that Bretz outlived every one of his prominent detractors as he survived almost to the age of ninety-nine. His sister once told me that her brother had been a man who could genuinely relish a good funeral.

The story of the long battle between J Harlen Bretz and the dogmatists of geology has been told a good many times. It seems customary for those who tell it to finish with a rousing homily about how this episode opened the eyes of geologists to the great world of catastrophic events, inspired them to explore new avenues in their thinking. I wish that were so, but know that it is not.

GLOSSARY

andesite. The common rock of volcanic chains. Andesite comes in various shades of gray and is commonly speckled with little white crystals of plagioclase feldspar.

anticline. An arch of stratified rock with the layers bent in opposing directions.

basalt. A common volcanic rock; the bedrock of the Columbia Basin. It tends to break into palisades of vertical columns. Most basalt is flat black, but that on the Columbia Basin is brownish black.

bedrock. Solid rock that has not moved.

Belt rocks. An enormous variety of distinctive sedimentary rock common to large areas of western Montana, northern Idaho, and southeastern British Columbia. Most are approximately 1 billion years old, give or take a couple of hundred million years.

coulee. A dry streambed or gully.

current-ripple. Ordinarily, one of the little ripples that appear in the sand on a streambed, generally a fraction of an inch high and as much as several inches from crest to crest. The giant ripples of Glacial Lake Missoula and the scablands are as much as 35 feet high and several hundred feet from crest to crest.

discharge. The volume of water flowing through a stream.

erratic rock. A rock somehow transported and dropped some distance from its original home.

gneiss. A metamorphic rock that generally resembles granite except in having a streaky grain, like wood.

granite. A coarsely granular igneous rock composed mainly of quartz and feldspar, along with a black mineral. The feldspar, and the rock, are generally either pink or very pale gray.

gulch filling. A deposit of sand and gravel swept into the slackwater of a small tributary valley by a flood pouring down the main valley.

hackly. Jagged.

hydraulic dam. A constriction that impounds water by taking water in faster on the upstream side than it releases it on the downstream side.

kolk. An extremely strong vortex that swirls around a more or less vertical axis in deep water that is flowing very fast. Kolks are capable of plucking boulders out of solid bedrock.

monocline. A simple bend in layered rock.

moraine. A mound, ridge, or other accumulation of sediment carried and deposited by a glacier. Moraines are an unstratified mix of clay, sand, gravel, and boulders intermingled, which is called glacial till.

mudstone. Mud hardened into solid rock.

Palouse Hills. The rolling countryside composed of old dunes of windblown dust in eastern Washington and western Idaho.

Palouse silt. A deep and richly fertile deposit of windblown dust that covers much of eastern Washington and western Idaho.

rock flour. Finely ground rock pulverized as rocks grind against each other within a moving glacier.

scabland. A tract of land in eastern Washington that is unfit for farming because it is scarred with dry stream channels, full of rock outcrops, or covered with coarse gravel.

stream terrace. A remnant of an old floodplain left high and dry as the stream eroded its bed to a new floodplain at a lower level.

talus. Rock debris at the base of a cliff, also known as slide rock. Talus falls from cliffs when water freezes and expands in fractures in the rock, breaking it into rubble.

valley floor. The more or less flat area of floodplain and stream terraces in the bottom of a valley.

valley wall. The hillslope that borders a stream valley.

varves. Alternating layers of light and dark sediment deposited on the floor of a glacial lake. The light layers are mostly rock flour liberated from melting ice in the summer. The dark layers are largely organic matter that settled to the lake bottom in winter. Each pair of light and dark layers is a record of one year.

Selected Bibliography

Alden, W. C. 1927. Channeled Scablands and the Spokane flood. *Washington Academy of Science Journal* 17 (8):203.

——. 1953. Physiography and glacial geology of western Montana and adjacent areas. *U.S. Geological Survey Professional Paper* 231.

Allen, J. E. 1932. Contributions to the structure, stratigraphy and petrography of the lower Columbia River Gorge. Master's thesis, University of Oregon.

——. 1958. *Columbia River Gorge, Portland to the Dalles: Guidebook for Field Trip Excursions.* Cordilleran Section, Geological Society of America.

——. 1984. *The Magnificent Gateway: A Layman's Guide to the Geology of the Columbia River Gorge.* Portland, Ore.: Timber Press.

Allen, J. E., and M. Burns. 1987. *Cataclysms on the Columbia.* Portland, Ore.: Timber Press.

Allison, I. S. 1932. Spokane flood south of Portland, Oregon. *Geological Society of America Bulletin* 43:133–34.

——. 1932. New version of the Spokane flood. *Geological Society of America Bulletin* 44:675–722.

——. 1935. Glacial erratics in the Willamette Valley. *Geological Society of America Bulletin* 46:605–32.

——. 1941. Flint's fill-hypothesis for channeled scabland. *Journal of Geology* 49:54–73.

——. 1978. Late Pleistocene sediments and floods in the Willamette Valley. Oregon Department of Geology and Mineral Industries, *Ore Bin* 40 (11 and 12):177–202.

Alt, D., and R. L. Chambers. 1970. Repetition of the Spokane flood. *Abstracts.* First meeting of the American Quaternary Association

(AMQUA), 28 August–1 September at Yellowstone Park and Montana State University, Bozeman, p. 1.

Atwater, Brian F. 1984. Periodic floods from Glacial Lake Missoula in the San Poil arm of the Glacial Lake Columbia, northeastern Washington. *Geology* 12 (8):464–67.

Baker, V. R. [1973a] 1981. Erosional forms and processes for the catastrophic Pleistocene Missoula floods in eastern Washington. In *Fluvial Geomorphology: Publications in Geomorphology,* edited by M. Morisawa, 123–48. London: George Allen and Unwin.

———. 1973b. Paleohydrology and sedimentology of Lake Missoula flooding in eastern Washington. *Geological Society of America Special Paper* 144.

———. 1977. Lake Missoula flooding and the channeled scabland. In *Geologic Excursions in the Pacific Northwest,* Guidebook for the Geological Society of America, 1977 Annual Meeting, edited by E. H. Brown and R. C. Ellis, 399–414. Bellingham: Western Washington University Press.

Baker, V. R., ed. 1981. Catastrophic flooding, the origin of the channeled scabland. *Benchmark Papers in Geology* 55. Stroudsburg, Penn.: [New York]: Dowden, Hutchinson, and Ross, Inc.

Baker, V. R., and D. J. Milton. 1974. Erosion by catastrophic floods on Mars and Earth. *Icarus* 23:27–41.

Baker, V. R., and D. Nummedal. 1978. *The channeled scabland: A guide to the geomorphology of the Columbia Basin, Washington.* Washington, D.C.: National Aeronautics and Space Administration.

Baker, V. R., G. Benito, and A. N. Rudoy. 1993. Paleohydrology of late Pleistocene superflooding, Altay Mountains, Siberia. *Science* 259:348.

Barnes, H. L. 1956. Cavitation as a geological agent. *American Journal of Science* 254:493–505.

Beatty, C. B. 1962. Topographic effects of Glacial Lake Missoula: a preliminary report. *California Geographer* 2:113–22.

Bretz, J H. 1923a. Glacial drainage on the Columbia Plateau. *Geological Society of America Bulletin* 34: 573–608.

———. 1923b. The Channeled Scabland of the Columbia Plateau. *Journal of Geology* 31:617–49.

———. 1924. The Dalles type of river channel. *Journal of Geology* 32:139–49.

————. 1925. The Spokane flood beyond the Channeled Scabland. *Journal of Geology* 33:97–115, 236–59.

————. 1927a. Channeled Scabland and the Spokane flood. *Washington Academy of Science Journal* 17(8):200–211.

————. 1927b. The Spokane flood: A reply. *Journal of Geology* 35:461–68.

————. 1928a. Alternative hypothesis for Channeled Scabland. *Journal of Geology* 36:193–223, 312–41.

————. 1928b. Bars of the Channeled Scabland. *Geological Society of America Bulletin* 39:643–702.

————. 1928c. The Channeled Scabland of eastern Washington. *Geographical Review* 18:446–77.

————. 1929. Valley deposits immediately east of the Channeled Scabland of Washington. *Journal of Geology* 37:393–427, 505–41.

————. 1930a. Lake Missoula and the Spokane flood. *Geological Society of America Bulletin* 41:92–3.

————. 1930b. Valley deposits immediately west of the Channeled Scabland. *Journal of Geology* 38:385–422.

————. 1932a. *The Channeled Scabland: 16th International Geological Congress, Guidebook 22.* Excursion C-2. Washington, D.C.: GPO.

————. 1932b. The Grand Coulee. *American Geographical Society Special Publication* 15. New York: American Geographical Society.

————. 1942. Lake Missoula and the Spokane flood. *Geological Society of America Bulletin* 53:1569–1600.

————. 1969. The Lake Missoula floods and the Channeled Scabland. *Journal of Geology* 77:503–43.

Bretz, J H., H. T. U. Smith, and G. E. Neff. 1956. Channeled Scabland of Washington: New data and interpretations. *Geological Society of America Bulletin* 67:957–1049.

Campbell, M., and others. 1916. Guidebook of the Western United States: Part A, The Northern Pacific Route. *U.S. Geological Survey Bulletin* 611.

Carson, R. J., C. F. McKhaun, and M. H. Pizey. 1978. The Touchet beds of the Walla Walla Valley. In *The channeled scabland: A guide to the geomorphology of the Columbia Basin*, edited by V. R. Baker and D. Nummedal, 173–77. Washington, D.C.: National Aeronautics and Space Administration.

Chamberlain, T. C. 1886. Administrative Report, *U.S. Geological Survey Seventh Annual Report, 1885–1886*. Washington, D.C.: GPO, p. 78.

Chambers, R. L. 1971. Sedimentation in Lake Missoula. Master's thesis, University of Montana.

Flint, R. F. 1935. Glacial features of the southern Okanogan region. *Geological Society of America Bulletin* 46:169–94.

————. 1936. Stratified drift and deglaciation of eastern Washington. *Geological Society of America Bulletin* 47:1849–84.

————. 1937. Pleistocene drift border in eastern Washington. *Geological Society of America Bulletin* 48:203–31

————. 1938a. Origin of the Cheney-Palouse Scabland tract. *Geological Society of America Bulletin* 49:461–524.

————. 1938b. Summary of the late-Cenozoic geology of southern Washington. *American Journal of Science* (5th Series) 35.

Flint, R. F. 1957. *Glacial and Pleistocene Geology*. New York: John Wiley and Sons.

Flint, R. F. and W. H. Irwin. 1939. Glacial geology of Grand Coulee Dam, Washington. *Geological Society of America Bulletin* 50:661–80.

Fryxell, R. 1962. A radiocarbon limiting date for scabland flooding. *Northwest Science* 36:113–19.

————. 1965. Mazama and Glacier Peak volcanic ash layers: relative ages. *Science* 147:1288–90.

Gilluly, J. C., 1927. Discussion: Channeled Scabland and the Spokane flood. *Washington Academy of Science Journal* 17(8)203–5.

Gilluly, J. C., A. C. Waters, and A. O. Woodford. 1968. *Principles of Geology*, Third Edition. San Francisco: W.H. Freeman and Company.

Glenn, J. L. 1965. Late Quaternary sedimentation and geologic history of the north Willamette Valley, Oregon. Ph.D. diss., University of Washington.

Hodge, E. T. 1931. Exceptional moraine-like deposits in Oregon. *Geological Society of America Bulletin* 42:985–1010.

————. 1934. Origin of the Washington scabland. *Northwest Science* 8:4–11.

————. 1950. The Belleview erratic. *Geological Society of the Oregon Country* 16(11):92–94

Kelly, Mary Pardee. 1963. Memorial to Joseph Thomas Pardee (1871–1960) *Geological Society of America Proceedings*, 39–41.

Lawrence, D. B., and E. G. Lawrence. 1958. Bridge of the Gods legend, its origin, history, and dating. *Mazama* 11(13):33–41.

Lewis, P. F. 1960. Linear topography in the southwestern Palouse, Washington-Oregon. *Annals of the American Association of Geographers* 50:98–111.

MacDonald, E. V., and A. J. Busacco. 1988. Record of pre-late Wisconsin giant floods in the Channeled Scabland interpreted from loess deposits. *Geology* 16:728–31.

Masursky, H., J. M. Boyce, A. L. Dial, G. G. Schaber, and M. E. Strobell. 1977. Classification and time of formation of Martian Channels based on Viking data. *Journal of Geophysical Research* 82:4016–38.

McKnight, E. T. 1927. The Spokane flood: A discussion. *Journal of Geology* 35:453–60.

Meinzer, O. E. 1916. Artesian water for irrigation in Little Bitterroot Valley, Montana. *U.S. Geological Survey Water Supply Paper 400 B.*

———. 1927. Discussion: Channeled scabland and the Spokane flood. *Washington Academy of Science Journal* 17(8):207–8.

Mullineaux, D. R., H. J. Hyde, and M. Rubin. 1975. Widespread glacial and post-glacial tephra deposits from Mt. St. Helens volcano, Washington. *U.S. Geological Survey Journal of Research* 3(3):329–35.

Mullineaux, D. R., R. E. Wilcox, W. F. Ebaugh, R. Fryxell, and M. Rubin. 1978. Age of the last major scabland flood of the Columbia River plateau in eastern Washington. *Quaternary Research* 10:171–8.

Nobles, L. H. 1952. Glacial sequence in the Mission Valley, Western Montana. *Geological Society of America Bulletin* 63:1286.

Palmer, L. 1977. Large landslides of the Columbia River Gorge, Oregon and Washington. *Geological Society of America, Reviews in Engineering Geology* 3:69–83.

Pardee, J. T. 1910. The Glacial Lake Missoula, Montana. *Journal of Geology* 18:376–86.

———. 1922. Glaciation in the Cordilleran region. *Science* 56:686–7.

———. 1942. Unusual currents in Glacial Lake Missoula, Montana. *Geological Society of America Bulletin* 53:1569–1600.

Parsons, R. B., and C. A. Balster. 1969. Late Pleistocene stratigraphy, southern Willamette Valley. *Northwest Science* 43(3):116–29.

Parsons, R. B., C. A. Balster, and A. O. Ness. 1970. Soil development and geomorphic surfaces, Willamette Valley, Oregon. *Soil Science Society of America, Proceedings* 34:485–9.

Patton, P. C., V. R. Baker, and R. C. Kochel. 1978. New evidence for pre-Wisconsin flooding in the channeled scabland of eastern Washington. *Geology* 6:567–71.

Rigby, J. G. 1982. The sedimentology, mineralogy, and depositional environment of a sequence of Quaternary catastrophic flood-derived lacustrine turbidites near Spokane, Washington. Master's thesis, University of Idaho.

Sharp, R. P., and M. C. Malin. 1975. Channels on Mars. *Geological Society of America Bulletin* 86:593–609.

Stauffer, J. 1956. Late Pleistocene flood deposits in the Portland area. *Geological Society of the Oregon Country Newsletter* 22(3):21–31.

Symons, T. W. 1882. *The Upper Columbia River and the Grand Plain of the Columbia.* 47th Congress, First Session, Senate Executive Document 186, 1–135.

Trimble, D. E. 1963. Geology of Portland, Oregon, and adjacent areas. *U.S. Geological Survey Bulletin* 1119.

U.S. Forest Service (Jeff Silkwood). 1998. *Glacial Lake Missoula and the Channeled Scabland: A digital portrait of landforms of the last Ice Age, Washington, Oregon, northern Idaho, western Montana.*

Waitt, R. B., Jr. 1977. Guidebook to Quaternary geology of the Columbia, Wenatchee, Peshastin, and upper Yakima valleys, west central Washington. *U.S. Geological Survey Open-File Report* 77-753.

————. 1980. About 40 last-glacial Lake Missoula jokulhlaups through southern Washington. *Journal of Geology* 88:653–79.

————. 1983. Tens of successive, colossal Missoula floods at north and east margins of Channeled Scabland. *U.S. Geological Survey Open-File Report* 83-671.

————. 1984. Periodic jokulhlaups from Pleistocene Glacial Lake Missoula—new evidence from varved sediment in northern Idaho and Washington. *Quaternary Research* 22:46–58.

————. 1985. Case for periodic, colossal jokulhlaups from Glacial Lake Missoula. *Geological Society of America Bulletin* 96:1271–86.

Walker, R. G. 1967. Varved lake beds in northern Idaho and northeastern Washington. *U.S. Geological Survey Professional Paper* 575-B:83–7.

Webster, G. D., V. R. Baker, and C. Gustafson. 1976. Channeled scabland of southeastern Washington, a roadlog via Spokane–Coulee City–Vantage–Washtucna–Lewiston–Pullman. *Field Guide No. 2, 72nd*

Annual Cordilleran Sectional Meeting, Geological Society of America. Pullman: Department of Geology, Washington State University.

Weis, P., and W. L. Newman. 1971. *The channeled scabland of eastern Washington: The geologic story of the Spokane flood.* U.S. Geological Survey pamphlet.

INDEX

ABOUT THE AUTHOR

David Alt, a geology professor at the University of Montana in Missoula, has studied Glacial Lake Missoula and its floods since the 1960s. Dedicated to bringing geology to the general public, he co-founded the popular Roadside Geology series. Alt also teaches Elderhostel courses, leads field trips, and presents public lectures about regional geology. He lives in Missoula, within view of the glacial lake's shorelines.